THE STATE OF DEVELOPMENT OF WASTE FORMS FOR MIXED WASTES

U.S. DEPARTMENT OF ENERGY'S OFFICE OF ENVIRONMENTAL MANAGEMENT

Committee on Mixed Wastes

Board on Radioactive Waste Management
Commission on Geosciences, Environment, and Resources
National Research Council

NATIONAL ACADEMY PRESS
Washington, D.C.

NOTICE: The project that is the subject of this report was approved by the Governing Board of the National Research Council, whose members are drawn from the councils of the National Academy of Sciences, the National Academy of Engineering, and the Institute of Medicine.

Support for this study was provided by the U.S. Department of Energy, under Grant No. DE-FC01-94EW54069. All opinions, findings, conclusions, and recommendations expressed herein are those of the authors and do not necessarily reflect the views of the Department of Energy.

International Standard Book Number 0-309-06431-7

Additional copies of this report are available from:

National Academy Press
2101 Constitution Avenue, N.W.
Box 285
Washington, DC 20055
800-624-6242
202-334-3313 (in the Washington Metropolitan Area)
http://www.nap.edu

Cover photo: Micrograph of a polymer encapsulated waste.
Photo courtesy of the Department of Energy's Mixed Waste Focus Area.

Printed in the United States of America.

COMMITTEE ON MIXED WASTES

PAUL A. DEJONGHE, *Chair,* Study Centre for Nuclear Research (retired), Mol, Belgium
ANN N. CLARKE, ANC Associates, Inc., Brentwood, Tennessee
JURGEN H. EXNER, JHE Technology Systems, Inc., Alamo, California
KENT F. HANSEN, Massachusetts Institute of Technology, Cambridge
JOANN S. LIGHTY, University of Utah, Salt Lake City
RICHARD J. SAMELSON, Consultant, Pittsburgh, Pennsylvania
MARTIN J. STEINDLER, Argonne National Laboratory (retired), Argonne, Illinois
BRUCE M. THOMSON, University of New Mexico, Albuquerque

Staff

JOHN R. WILEY, Senior Staff Officer (January 1998 to present)
KARYANIL THOMAS, Senior Staff Officer (through December 1997)
ANGELA R. TAYLOR, Senior Project Assistant
LAURA D. LLANOS, Project Assistant

BOARD ON RADIOACTIVE WASTE MANAGEMENT

MICHAEL C. KAVANAUGH, *Chair*, Malcolm Pirnie, Inc., Oakland, California
JOHN F. AHEARNE, *Vice-Chair*, Sigma Xi and Duke University, Research Triangle Park, North Carolina
ROBERT J. BUDNITZ, Future Resources Associates, Inc., Berkeley, California
MARY R. ENGLISH, University of Tennessee, Knoxville
DARLEANE C. HOFFMAN, Lawrence Berkeley National Laboratory, Oakland, California
JAMES H. JOHNSON, JR., Howard University, Washington, D.C.
ROGER E. KASPERSON, Clark University, Worcester, Massachusetts
JAMES O. LECKIE, Stanford University, Stanford, California
JANE C. S. LONG, Mackay School of Mines, University of Nevada, Reno
CHARLES MCCOMBIE, Consultant, Gipf-Oberfrick, Switzerland
WILLIAM A. MILLS, Oak Ridge Associated Universities (retired), Olney, Maryland
D. WARNER NORTH, NorthWorks, Inc., Mountain View, California
MARTIN J. STEINDLER, Argonne National Laboratories (retired), Argonne, Illinois
JOHN J. TAYLOR, Electric Power Research Institute (retired), Palo Alto, California
MARY LOU ZOBACK, U.S. Geological Survey, Menlo Park, California

Staff

KEVIN D. CROWLEY, Director
ROBERT S. ANDREWS, Senior Staff Officer
THOMAS KIESS, Senior Staff Officer
JOHN R. WILEY, Senior Staff Officer
SUSAN B. MOCKLER, Research Associate
TONI GREENLEAF, Administrative Assistant
PATRICIA A. JONES, Senior Project Assistant
ANGELA R. TAYLOR, Senior Project Assistant
LATRICIA C. BAILEY, Project Assistant
MATTHEW BAXTER-PARROTT, Project Assistant
LAURA D. LLANOS, Project Assistant

COMMISSION ON GEOSCIENCES, ENVIRONMENT, AND RESOURCES

The National Academy of Sciences is a private, nonprofit, self-perpetuating society of distinguished scholars engaged in scientific and engineering, research, dedicated to the furtherance of science and technology and to their use for the general welfare. Upon the authority of the charter granted to it by the Congress in 1863, the Academy has a mandate that requires it to advise the federal government on scientific and technical matters. Dr. Bruce Alberts is president of the National Academy of Sciences.

The National Academy of Engineering was established in 1964, under the charter of the National Academy of Sciences, as a parallel organization of outstanding engineers. It is autonomous in its administration and in the selection of its members, sharing with the National Academy of Sciences the responsibility for advising the federal government. The National Academy of Engineering also sponsors engineering programs aimed at meeting national needs, encourages education and research, and recognizes the superior achievements of engineers. Dr. William A. Wulf is president of the National Academy of Engineering.

The Institute of Medicine was established in 1970 by the National Academy of Sciences to secure the services of eminent members of appropriate professions in the examination of policy matters pertaining to the health of the public. The Institute acts under the responsibility given to the National Academy of Sciences by its congressional charter to be an adviser to the federal government and, upon its own initiative, to identify issues of medical care, research, and education. Dr. Kenneth Shine is president of the Institute of Medicine.

The National Research Council was organized by the National Academy of Sciences in 1916 to associate the broad community of science and technology with the Academy's purposes of furthering knowledge and of advising the federal government. Functioning in accordance with general policies determined by the Academy, the Council has become the principal operating agency of both the National Academy of Sciences and the National Academy of Engineering in providing services to the government, the public, and the scientific and engineering communities. The Council is administered jointly by both Academies and the Institute of Medicine. Dr. Bruce Alberts and Dr. William A. Wulf are chairman and vice-chairman, respectively, of the National Research Council.

Acknowledgement of Reviewers

This report has been reviewed by individuals chosen for their diverse perspectives and technical expertise, in accordance with procedures approved by the National Research Council (NRC) Report Review Committee. The purpose of this independent review is to provide candid and critical comments that will assist the NRC in making the published report as sound as possible and to ensure that the report meets institutional standards for objectivity, evidence, and responsiveness to the study charge. The content of the review comments and draft manuscript remain confidential to protect the integrity of the deliberative process. We wish to thank the following individuals, who are neither officials nor employees of the NRC, for their participation in the review of this report:

Bill Batchelor, Texas A&M University
Robert M. Bernero, U.S. Nuclear Regulatory Commission (retired)
Elizabeth M. Drake, Massachusetts Institute of Technology
Rodney C. Ewing, University of Michigan
Michael C. Kavanaugh, Malcolm Pirnie, Inc.
William A. Mills, Oak Ridge Associated Universities (retired)
Chris W. Rhyne, U.S. Environmental Protection Agency
Roger R. Seitz, International Atomic Energy Agency
E-an Zen, University of Maryland

While the individuals listed above have provided many constructive comments and suggestions, it must be emphasized that responsibility for the final content of this report rests entirely with the authoring committee and the NRC.

Contents

Executive Summary

More than 167,000 cubic meters of mixed waste, waste that contains both chemically hazardous and radioactive components, are in the known inventory at U.S. Department of Energy (DOE) sites that formerly produced nuclear defense materials. The inventory includes both mixed low-level wastes (MLLW) and mixed transuranic (MTRU) wastes.[1] Site cleanup and decommissioning activities during the next several years are expected to nearly double this inventory, and the inventory will be further increased by mixed wastes retrieved as a result of DOE site remediation.[2] Processing and permanent disposal of these mixed wastes is a part of the DOE Office of Environmental Management (EM) program to close former DOE production sites (DOE, 1998c).

---

[1] Hazardous wastes are defined by the Resource Conservation and Recovery Act (RCRA) of 1976. Radioactive materials are defined by the Atomic Energy Act of 1954. Only MLLW and MTRU are dealt with in this report. Mixed high-level radioactive wastes are not included because their radiation hazard requires an infrastructure for their regulation, treatment, and disposal that is not applicable to MLLW and MTRU. Mixed wastes from uranium mining and milling are also excluded.

[2] Eighteen DOE sites include waste disposal areas that must be remediated under the Comprehensive Environmental Response, Compensation, and Liability Act (CERCLA) of 1980. CERCLA is often referred to as the Superfund act. Site remediation is not expected to generate wastes that must be handled or processed differently from those in the current inventory.

Within EM, the Office of Science and Technology (OST, EM-50) is charged with assuring that safe, cost-effective technologies are available for the entire closure program. To address mixed waste technology needs, OST established a special program and management team, the Mixed Waste Characterization, Treatment, and Disposal Focus Area (MWFA), in 1994. At the request of OST, the National Research Council (NRC) convened a Committee on Mixed Waste[3] to assess specific technical issues being addressed by the MWFA. A review of all OST technology development activities was completed in 1995 (NRC, 1996c).

For the present task, the mixed waste committee was requested to review and evaluate the state of development of the final forms for disposal of mixed wastes as they arise from current and emerging treatment technologies.[4] The review was also to identify the technology development options DOE might consider in order to achieve waste forms that are cost effective and safe for disposal.[5]

In carrying out the review, the committee received formal presentations from DOE staff and other individuals, and examined documents and data provided by DOE and other sources. Some committee members visited waste contractors and DOE sites to gather additional information. The committee assessed the state of development of waste forms within the context of DOE's site closure program, technical approaches, and constraints (DOE, 1998c). The following subjects are discussed and commented upon in this report:

- the current and expected inventory of DOE mixed waste;
- laws and regulations that control mixed waste management;

---

[3] To be referred to as the mixed waste committee or the committee throughout this report. This committee is successor to a subcommittee (of the same name) of the Committee on Environmental Management Technologies (CEMT). The CEMT subcommittees were reorganized as independent committees under the National Research Council's Board on Radioactive Waste Management in 1997.

[4] The committee's Statement of Task is in Appendix A.

[5] Depending on its nature, mixed waste can be disposed of either in the form in which it was generated or, more often, after treatment to render the waste suitable for disposal. For the purpose of this report, a "waste form" is considered to be a solid material that is the product of one or more treatment processes.

- technologies for treating mixed wastes and producing the waste forms;
- tests to characterize the chemical and physical durability of waste forms; and
- methods to demonstrate the long-term performance of the disposal system.

The committee's general finding is that currently available waste forms are adequate (sufficiently developed) to meet regulatory requirements for disposal of DOE's known and expected mixed waste inventory. The general classes of waste forms that are available to the MWFA include grout, glass, ceramics, polymers, and compacted waste. Many of these waste forms have resulted from the intensive worldwide efforts and experience in developing waste forms for high- and low-level radioactive waste. No single form is appropriate for all wastes, but collectively the variety of available waste forms and well-established waste form production technologies make it unlikely that any totally new class of waste forms will be necessary to complete EM's planned cleanup program.

There are a number of caveats to this finding:

1. Analysis by DOE concludes that nearly all of EM MLLW can be safely disposed at Hanford, Washington and Envirocare in Utah due to their dry climates. However, it is likely that political and economic considerations will establish a need for other disposal facilities. Uncertainties regarding where these facilities will be located and future waste acceptance criteria introduce significant risk in judging the adequacy of EM's planned mixed waste treatment and stabilization processes.
2. Optimization of existing technologies to allow higher concentrations of waste in the product waste forms and provide less expensive production methods is possible, and potential cost savings justify continued effort toward process optimization. In particular, fabricating waste forms that contain higher concentrations of ash and salt would reduce the volume of waste to be disposed and the concomitant disposal costs. Methods to stabilize mercury need optimization.

3. There are no generally accepted tests for evaluating the integrity of any waste form over the required 1,000- to 10,000-year time frame. Consequently, performance assessments take little or no credit beyond a few hundred years for the waste form, and rely on other features of the overall disposal system to demonstrate compliance with long-term release requirements.[6]

The committee found that the waste form is not a "stand alone" entity and must be part of an integrated systems approach to mixed waste management. This approach includes the waste itself, regulations, treatment technologies, characterization of the waste form, and performance assessment of the disposal system. Within this systems approach there are several specific areas that may be the source of future problems. These include the following:

1. EM has a good qualitative knowledge of its mixed waste inventory. However, quantitative data that are necessary for developing reliable, cost-effective flowsheets and process optimization are deficient.
2. Proposed treatment technologies for mixed waste are based on technologies developed for sanitary, hazardous, or radioactive wastes. Actual experience in engineering adaptation, process integration, and operation for DOE's various mixed waste streams is lacking.
3. Other than for the few existing sites, there is no detailed site information for low-level or mixed waste disposal facilities. Demands that geology, hydrology, future demography, or intrusion will place on the waste form are often speculative.
4. Regulations controlling mixed wastes are imposed by several agencies and are complex, confusing, and subject to change and interpretation. Demonstrating compliance is a difficult and moving target.

In view of its general and specific findings, the committee makes the following recommendations regarding the state of development of waste forms for mixed waste and possible future directions for OST's mixed waste program:

---

[6] Performance assessment modeling is discussed in Chapter 6.

1. Waste form development should no longer be a primary focus of MWFA. Future research and development should emphasize integration, demonstration, and optimization of performance and reliability of mixed waste treatment processes, cost reduction, and provision of a more quantitative description of EM's mixed waste inventory.

2. MWFA should continue its practice of defining, identifying, and responding to technology deficiencies. The MWFA has established a rational and systematic program that identifies and prioritizes deficiencies. The committee compliments MWFA on this effort and encourages continued updates of the Technical Baseline Report that documents the state of its technology development activities.

3. MWFA should broaden its use of a systems approach in its efforts to assist EM in its closure program. This approach includes determination of the characteristics of the raw waste, definition of the required performance of a proposed technology, and design of the technology to attain the required performance. An important aspect of a reliable system design is flexibility to accommodate new information, experience, and reasonable changes in the performance requirements.

4. EM should work with the U.S. Environmental Protection Agency and the U.S. Nuclear Regulatory Commission to agree on clear guidelines that describe acceptable waste forms for disposal of mixed waste in future, near-surface disposal facilities. This should be done as soon as possible to reduce the risk that EM will deploy technologies that are later judged to be inadequate.

1

Introduction

More than 167,000 cubic meters of mixed waste, waste that contains both chemically hazardous and radioactive components, are in the known inventory at U.S. Department of Energy (DOE) sites that formerly produced nuclear defense materials. The inventory contains low-level radioactive waste mixed with hazardous materials, referred to as mixed low-level waste (MLLW), and transuranic waste mixed with hazardous materials, referred to as mixed transuranic (MTRU) waste.[1] Site cleanup and decommissioning activities during the next several years are expected to nearly double this inventory. The inventory will be further increased by mixed wastes retrieved as a result of DOE site remediation.[2]

Processing and permanent disposal of these mixed wastes is a part of the DOE Office of Environmental Management (EM) program to

---

[1] Hazardous wastes are defined by the Resource Conservation and Recovery Act of 1976. Radioactive materials are defined by the Atomic Energy Act of 1954. Only MLLW and MTRU are dealt with in this report. Mixed high-level radioactive wastes are not included because their radiation hazard requires an infrastructure for their regulation, treatment, and disposal that is not applicable to MLLW and MTRU. Mixed wastes from uranium mining and milling are also excluded.

[2] Eighteen DOE sites include waste disposal areas that must be remediated under the Comprehensive Environmental Response, Compensation, and Liability Act (CERCLA) of 1980. CERCLA is often referred to as the Superfund act.

close former DOE production sites. Within EM, the Office of Science and Technology (OST, EM-50) is charged with assuring that technologies are available to support the EM cleanup program (DOE, 1998c). OST's Mixed Waste Focus Area (MWFA) is responsible for identifying and developing technologies for processing EM's mixed wastes. The final, solid form of the waste that results from waste processing and is intended for disposal is referred to as the waste form. The waste form is a key part of an integrated waste management system.

Mixed waste management and disposal are complex technical and regulatory challenges. It is a technical challenge because of both the quantity and wide variety of mixed wastes within EM's responsibility. Examples range from small amounts of laboratory wastes that contain many chemicals and relatively high levels of radioactivity, to large volumes of soil and debris that contain only slightly contaminated material. Treatment technologies, therefore, must deal with a broad range of chemical and physical properties and volumes. Treatment objectives are generally to reduce the chemical hazards of a given waste material and to render it into a stable waste form. The long-term durability of the waste form must be assured to make it acceptable for disposal.

Regulatory complexities arise because regulations have been developed by two different regulatory agencies that have taken two different approaches to protecting the environment. Regulations apply to handling, storing, transporting, and treating the waste, to characteristics of the waste form, and to the disposal facility. The U.S. Nuclear Regulatory Commission has authority to license low-level waste disposal sites. The U.S. Environmental Protection Agency has authority to regulate hazardous waste, which includes discarded materials that are corrosive, ignitable, reactive, or contain toxic compounds and metals.[3]

This report is the result of a one-year review of the status of waste forms for the disposal of mixed wastes under the responsibility of EM. The review was performed by the Committee on Mixed Waste of the National Research Council's Board on Radioactive Waste Management (BRWM) at the request of OST. In its statement of task, the committee

---

[3] Although the word "metals" is used for convenience throughout this report, it is recognized that the chemical forms of these metals that exist in actual wastes include oxides, salts, and complexes. The zero valent state is not to be assumed.

was requested to review and evaluate the state of development of the final forms of treated waste; to assess the characteristics (and their uncertainties) of the waste form for disposal; and identify requirements for additional research and development.[4] The committee responded to the statement of task by, first of all, putting its evaluation of the state of development of waste forms into the context of EM's mixed waste inventory and the regulations that control treatment and disposal of this waste. This context is provided in Chapters 2 and 3. Technologies to treat the waste inventory and produce waste forms are described in Chapter 4. Characterization of waste forms is described in Chapter 5, and Chapter 6 describes the performance assessment (PA) methodology to evaluate the long-term safety of disposal. The committee's findings and recommendations in Chapter 7 identify needs for additional research and development. The results of this study are intended to help OST consider options in technology development to achieve waste forms that are cost-effective and safe for disposal.

In carrying out the review, the committee heard formal presentations from DOE staff and other individuals and examined reports and data provided by DOE or other sources. Some committee members visited waste contractors and DOE sites to gather additional information. Because waste forms are part of an integrated waste management system, the committee evaluated their state of development in the context of EM's cleanup program and current regulatory requirements. The information considered by the committee can be divided into the following categories:

- scope of the mixed waste problem, including the waste inventory and characteristics;
- the MWFA approach and activities, including technology development and application;
- performance of selected waste forms in laboratory testing, as well as evaluation of specific waste forms at potential disposal sites; and
- regulations and waste acceptance criteria.

---

[4] The Statement of Task is in Appendix A.

A synopsis of the information presented to the committee is given in Appendix B.

HISTORY OF THE MWFA AND THE MIXED WASTE COMMITTEE

The DOE established its Office of Environmental Management (EM) in November 1989. EM's mission was to reduce threats to health and safety posed by contamination and waste at DOE sites (DOE, 1998c). Within EM, OST[5] was created to promote the development of new and improved technologies needed to effect remediation, to lower cleanup costs, and to reduce risks.

In 1994, OST reorganized its work with the creation of a new management structure to focus its efforts on EM's most urgent environmental restoration and waste management problems. The following focus areas were established:

1. high-level radioactive waste tank remediation,
2. mixed waste characterization, treatment, and disposal,
3. contaminant plume containment and remediation,
4. landfill stabilization,[6] and
5. facility transitioning, decommissioning, and final disposition.[7]

In 1994, as an adjunct to this new structure, the DOE Assistant Secretary for EM requested the BRWM to form a Committee on Environmental Management Technologies (CEMT) to provide independent reviews of OST programs and to give recommendations on technology development and use. As part of this effort, a Subcommittee on Mixed Waste Characterization, Treatment, and Disposal was formed under the CEMT. The mixed waste subcommittee, as well as the other

[5] OST was previously named the Office of Technology Development.

[6] Focus areas 3 and 4 were combined and renamed Subsurface Contaminants.

[7] First renamed Decontamination and Decommissioning, and more recently Deactivation and Decommissioning (DOE, 1998d).

CEMT subcommittees, were reorganized as independent committees under the BRWM in 1997. A CEMT report that summarized findings from all five subcommittees was published in 1996 (NRC, 1996c).

ROLE OF THE WASTE FORM IN DOE MIXED WASTE MANAGEMENT

Mixed waste can be disposed of either in the form in which it originally was generated or, in most cases, after treatment to reduce volume and render the waste suitable for disposal. Knowledge of the inventory and characteristics of the waste itself is necessary to determine the subsequent steps for its treatment and disposal. For mixed wastes within the responsibility of EM, treatments are being identified and developed by the MWFA that are designed to reduce waste volume, destroy hazardous organic materials, remove or stabilize toxic chemicals, and produce waste forms that meet disposal criteria (DOE, 1997a, 1996a). A waste form is considered to be a solid material that is the product of one or more treatment processes. During interim storage and after disposal, the waste form should constitute an important barrier against dispersion of hazardous and radioactive components.

For disposal of waste in a near-surface facility,[8] protection of the environment can be achieved by a complete disposal system that includes the following in its design:

- the waste form;
- additional engineered barriers such as synthetic and clay liners that are intended to retard migration of contaminants from the repository;
- site geology, which affects the contaminant migration beyond the engineered barriers;
- ground water movement, which affects the time and concentration at which contaminants become a concern; and

[8] Near-surface disposal is a standard practice for hazardous or low-level radioactive waste. The specially designed and constructed facility is typically located at least 3 m, but no more than about 30 m, below the surface of the surrounding land, and usually above the water table.

- climatic conditions.

The waste form should not be considered a separate entity; rather, it is an integral part of the entire treatment and disposal system.

The primary function of a waste form is to retain the hazardous and radioactive constituents of the waste, or to retard their release.[9] In addition, the physical stability of the waste form facilitates its handling, storage, and transportation. For disposal in a near-surface facility, the mechanical strength and stability of the waste form will also help to prevent subsidence and assist in maintaining the integrity of the cover. In case of inadvertent intrusion into the repository, a stable, high-integrity waste form can limit potential exposure by discouraging excavation or drilling, and by reducing the dispersal of the waste if the repository is breached (Berry, 1994).

On the other hand, waste forms that degrade rapidly through mechanical, chemical, biochemical, radiological, or other mechanisms may result in untimely release of contaminants and in some cases may actually facilitate migration of contaminants from the disposal facility. For example, poor quality waste forms can contribute to subsidence or cover failure which, in turn, increases the potential for release of contaminants into ground or surface water.

To determine whether a waste form is appropriate for disposal, waste form performance criteria must be identified. Compliance with the criteria can be demonstrated by characterization of the chemical and physical properties of the waste form, and testing its performance under laboratory and realistic disposal conditions. Reliable tests and testing protocols must be established by technology developers and regulators to provide a basis for waste form development and to judge if a proposed waste form is acceptable for a given waste and disposal condition.

---

[9] The useful life expectancy of presently available waste forms ranges from a few hundred to several thousand years. The waste form and engineered barriers can be expected to effectively retain radionuclides with half-lives of up to 30 years, e.g., ^{90}Sr and ^{137}Cs. In instances where the waste form and engineered barriers may degrade more rapidly than the contaminants in the waste, the entire disposal system is relied upon. Methods to measure durability are described in Chapter 5.

Mathematical models, known as performance assessment models, are used to describe the performance of the integrated waste disposal system over very long time periods. These PA models quantify the importance of individual components as well as the multiple components of the repository system including the stabilized waste form, natural and engineered containment systems associated with the disposal facility, and the disposal site geology to assure future protection of humans and the environment.

In reviewing and evaluating the state of development of waste forms, this report will present information gathered by the mixed waste committee and the committee's findings and recommendations in the context of mixed waste management. This includes the characteristics and inventory of EM's mixed wastes, regulatory controls, technology availability, characterization of waste forms, and use of PA models.

2

Inventory and Characteristics of DOE Mixed Waste

To support the selection of treatment technology and waste forms, the inventory and characteristics of stored waste material must be known with a reasonable degree of confidence. This section reviews documentation provided by the Mixed Waste Focus Area (MWFA) and discusses the adequacy of the present knowledge about the inventory of mixed waste under responsibility of the Department of Energy (DOE) Office of Environmental Management (EM). Two sources supplied most of the inventory information: the National 1995 Mixed Waste Inventory Report (MWIR) (DOE, 1995a) and the Mixed Waste Focus Area Technical Baseline Report (DOE, 1996a, 1997a).

The MWFA estimates that the quantity of mixed low-level waste (MLLW) and mixed transuranic (MTRU) waste presently in EM's inventory is 167,000 cubic meters (m^3).[1] Approximately two thirds of the waste is MLLW and the remainder is MTRU. Projections of mixed waste generation during the next few years indicate that the inventory will increase to about 250,000 m^3. This does not include the waste that will require treatment as a result of remediation of closed sites under the Comprehensive Environmental Response, Compensation, and Liability

[1] The measure of waste by volume rather than weight is industry practice. All waste quantities and percentages given in this report are volumetric.

Act [2] regulations, which will add significantly to the total waste volume. While these additions will increase the total inventory, the current inventory is believed to be representative of all the various types of mixed wastes that must be treated and stabilized (Kolts, 1996). [3]

In its current inventory, EM distinguishes the following waste groups: (1) waste waters, (2) combustible organic waste, (3) homogeneous solids, sludges, and soils, (4) debris, and (5) unique wastes. The following page gives a list of these groups and describes the materials that comprise each group. To characterize individual waste types, the MWIR-1995 database uses 119 codes (DOE, 1995a). These codes cover acidic aqueous liquids, pond and other sludges, waste solvents, discarded equipment, discarded protective clothing, paint wastes, used air filters, discarded chemicals, lead shielding, contaminated soil, debris, and many other categories. The waste is stored in containers such as large storage tanks, 200 liter drums, boxes (up to 4 m^3), and numerous small containers of various shapes, sizes, and methods of construction. The containers are made from various types of wood, metal and plastic, but the specific container materials are not included in MWIR-1995.

The inventory includes wastes from 45 locations, however the following seven locations account for 96% of the total (DOE, 1995a). The locations of these seven sites are shown on Figure 1.

- Idaho National Engineering and Environmental Laboratory – 35% (includes much of the MTRU waste);
- Oak Ridge, including the laboratory, K-25 site and Y-12 plant – 30%;
- Rocky Flats Plant – 9%;

---

[2] Enacted by Congress in 1980 to require cleanup of sites contaminated by past activities, including "closed" sites where contamination by hazardous and toxic chemicals exists.

[3] Present-day operating practice applies the concept of waste minimization to production processes. In addition to process changes to reduce the quantity of waste generated, this concept includes segregation of wastes to simplify treatment and allow recycle of useful materials. Waste minimization is required by DOE order 5820.2A that is described in Chapter 3. It is expected that current and future wastes will be easier to treat than the older wastes, which as noted in this report, are heterogeneous and inadequately characterized.

List of Mixed Waste Groups that Comprise the EM Inventory

Waste Waters
Aqueous Liquids and Slurries.
This group includes waste waters and slurries having acidic, basic, and neutral characteristics, as well as cyanide-containing waste waters and slurries.

Combustible Organics
Organic Liquids
This group includes aqueous streams containing both halogenated and nonhalogenated organic compounds, as well as pure organic streams containing halogenated and nonhalogenated compounds.
Organic Homogeneous Solids
Organic particulate matter (resins, organic absorbents), organic sludges (biological sludges, halogenated and nonhalogenated sludges), and organic chemicals are included in this group.

Inorganic Homogeneous Solids and Soils
Inorganic Homogeneous Solids
These wastes include particulate matter—ash, sandblasting media, inorganic particulate absorbents, absorbed organic liquids, ion-exchange media, metal chips/turnings, glass/ceramic materials, and activated carbon.
Inorganic Sludges
Waste water treatment sludges, pond sludges, off-gas treatment sludges, plating waste sludges, and reprocessing sludges constitute this group.
Other Inorganic Waste
This group includes paint waste (paint chips and solids, paint sludges), salt waste containing chlorides, sulfates, nitrates, and metal oxides/hydroxides, and inorganic chemicals.
Solidified Homogeneous Solids
This group includes soil, soil/debris, and rock/gravel.

Debris
Metal Debris
Metal debris with or without lead or cadmium constitute this category.
Inorganic Nonmetal Debris
Concrete, glass, ceramic/brick, rock, asbestos, and graphite debris are included in this category.
Organic Debris
This category includes plastic/rubber, leaded gloves/aprons, halogenated plastics, nonhalogenated plastics, wood, paper, and biological debris.
Heterogeneous Debris
Composite filters, inorganic and organic debris, asphalt, and electronic equipment constitute this category.

Unique Waste
Lab Packs
Organic, aqueous, and solid lab packs and scintillation cocktails are included.
Special Wastes
Included in this category are elemental mercury, elemental hazardous metals (activated and non-activated lead, elemental cadmium), beryllium dust, batteries (lead acid, mercury, cadmium), reactive metals (bulk and reactive metal-contaminated components), pyrophoric fines explosives/propellants, and compressed gases/aerosols.
All others
Materials placed in a final waste form are included in this category.

SOURCE: (DOE, 1995a)

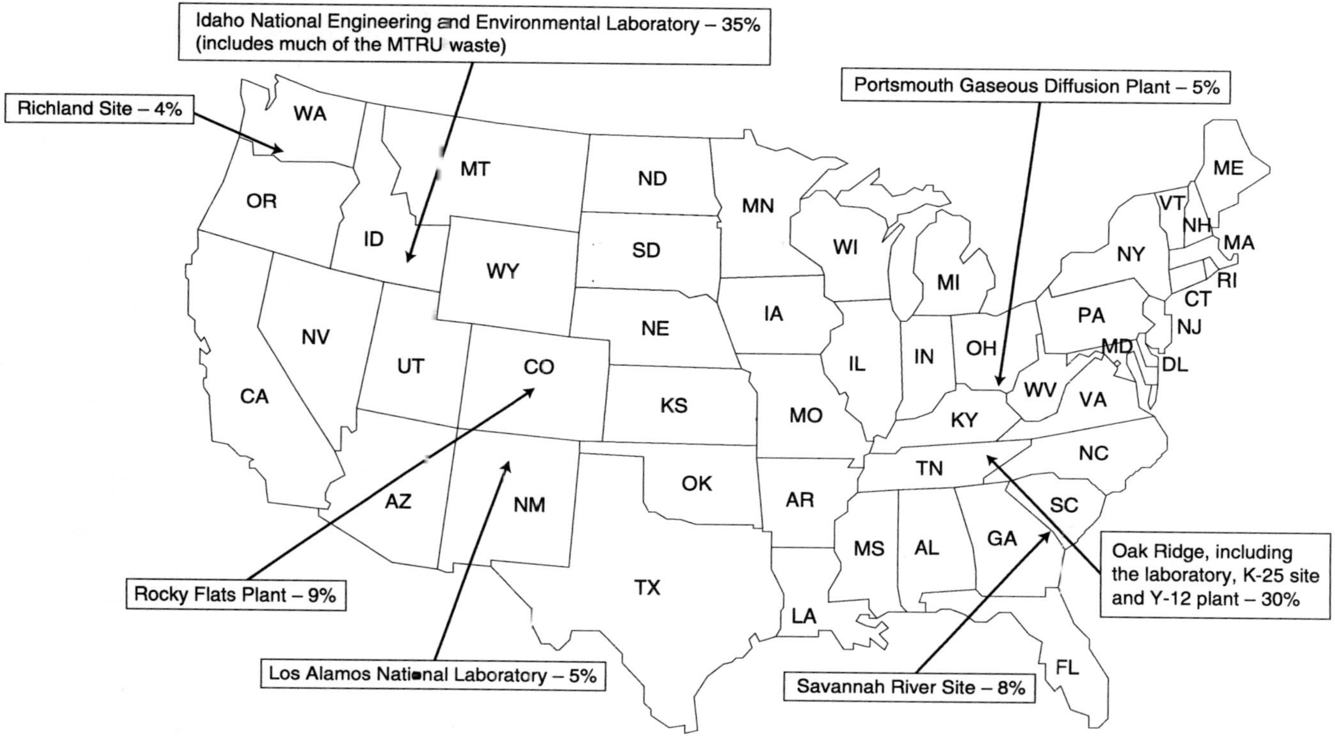

FIGURE 1 The seven locations where 96% of the total mixed waste inventory is stored. SOURCE: DOE, 1995a.

- Savannah River Site – 8%;
- Los Alamos National Laboratory – 5%;
- Portsmouth Gaseous Diffusion Plant – 5%;
- Richland Site – 4%.

The MWFA has identified 58 radionuclides that account for most of the radioactivity in the present inventory of EM's mixed wastes. Because the shorter-lived radionuclides have mostly decayed, those remaining range from ^{60}Co having a half-life of 5.27 years to ^{232}Th having a half-life of 14 billion years. While the radionuclide content varies widely among waste streams, the available data indicate that it is unusual for more than about six or eight nuclides to be present in any individual stream.

In accord with the U.S. Environmental Protection Agency (EPA) regulations, much of EM's waste inventory has been declared hazardous based on the processes that produced it, or other historical knowledge, but actual analyses are generally not available. Data from MWIR-1995 show that more than 80% of the inventory has been declared hazardous because it is suspected to be contaminated with waste solvents, electroplating and metal treating wastes, or waste from leachate treatment (DOE, 1995a). Solvent contamination is believed to occur in two thirds of the inventory. Lead contamination is suspected in more than half the inventory. Wastes regulated by the Toxic Substance Control Act (TSCA), such as polychlorinated biphenyls (PCBs), are suspected in about 6% of the mixed waste inventory, but may be present in up to half of the combustible organic waste.[4] TSCA-regulated combustible wastes can be burned only in incinerators with a TSCA permit to operate.

In addition to the available inventory data contained in MWIR-1995 (DOE, 1995a), the database includes a qualitative description of MWFA's confidence in the data. The confidence levels for the waste type, and the hazardous and radioactive waste components are presented as "high," "medium," "low" or, in a few cases, not available. Overall the confidence level in the type of waste is rated medium or high for about 84% of the inventory. Confidence in the radioactive component

---

[4] The Toxic Substances Control Act (TSCA) includes special management provisions for handling and cleaning up materials that are considered to present imminent hazards. TSCA is described in Chapter 3.

characterization is medium or high for 64% of the inventory. In the absence of actual analyses for most of the hazardous components, high or medium confidence in their characterization drops to about 29% of the overall inventory. The portion of the inventory for which confidence levels on both hazardous and radioactive components is medium or high is only about 25%. In the remaining 75%, confidence in one or both components is rated as low (or not available).

In spite of the uncertainties for the hazardous components in the inventory, the percent of the inventory that displays EPA hazardous characteristics can be estimated from MWIR-1995. Table 1 summarizes the hazardous characteristics of the entire inventory. Almost 60% of the inventory displays the toxicity characteristic, essentially all of which is due to heavy metals (58%).[5] Lead is present is most of this inventory (54%). Mercury contamination, which is difficult to treat as will be discussed in Chapter 4, is present in about 26% of the inventory. It should be emphasized, however, that the confidence level in the hazardous component characterization is generally low.

Possible measures to reduce the amount of waste that must be treated as mixed waste, such as better waste characterization and the establishment of "below regulatory concern" criteria are described in the committee's findings and recommendations in the following section. The EPA's procedure for de-listing waste and its proposed Hazardous Waste Identification Rule are mentioned in Chapter 3.

FINDINGS, DISCUSSION, AND RECOMMENDATIONS

The committee found that EM's mixed waste inventory is sufficiently characterized that conceptual design of treatment processes and waste form selection can proceed. However, the inventory is insufficiently characterized for detailed engineering design of treatment processes or process optimization. Better characterization is necessary to reduce risks of technology failures and allow cost-effective design and operation of treatment processes.

---

[5] The Toxicity Characteristic Leaching Procedure is discussed in Ch 3.

TABLE 1 Hazardous Characteristics as a Percent of the EM Mixed Waste Inventory

Hazardous Characteristic	Percent of Inventory Volume
Ignitable	22
Corrosive	21
Reactive	7
Toxic Chemicals	59
Heavy Metals	58
Mercury	26
Lead	54
Chromium	31
Cadmium	27
Organics	30
Non-specific Listed Contamination	82
Solvents	64
Electroplating Waste	34
Cyanide	15
Leachate Treatment	11

SOURCE: DOE (1995a).

The mixed waste inventory is stored in a variety of forms and conditions. Inventory diversity results from past waste management practices (e.g., poor segregation of wastes, and discharge to ponds or retention basins that now must be remediated) and past storage practices (e.g., drums of material having only generic identification of contents). The MWFA is confronted with complex, often poorly defined mixtures of waste materials.[6]

[6] The committee noted that at present there is no lower limit for contamination by radioactive nuclides below which the waste would simply be considered as hazardous, i.e., not mixed. Definition of *de minimis*, below regulatory concern, and exemption levels has been discussed by U.S. and international regulatory agencies for many years (IAEA, 1996). EPA is proposing a new risk-based "Hazardous Waste Identification Rule" that would allow waste containing only small amounts of hazardous waste to be removed from its system for regulating hazardous wastes. Although the committee did not receive information that would allow it to quantify the effects of such definitions, it appears that proper selection of lower limits for contamination could reduce the total volume of material considered as mixed waste, resulting in considerable cost savings.

Although qualitative information sufficient, in the committee's view, to support conceptual design of treatment processes is available for most mixed waste streams, quantitative analyses are incomplete, as discussed in this chapter. Considerable effort will be required to quantify the radiological and hazardous constituents in many waste streams. MWFA appears to have reasonably good descriptive knowledge about the waste types (e.g., paint waste or pond sludge), and fair knowledge about its radioactivity characteristics (e.g., nuclides present and radiation levels), but MWFA is lacking specific analytical knowledge about the hazardous chemical content of the waste.

In the absence of chemical analyses, much of EM's waste has been declared hazardous based on the process that produced it or other historical knowledge. Hazardous waste requires special, licensed facilities for treatment, storage and disposal (see Chapter 3) that significantly increase costs. Actual analyses may show that some of this declared inventory is not hazardous and thus reduce overall costs.

The MWFA has recognized the need for improved waste characterization by listing it as the number one priority on its list of technology needs.[7] The committee agrees that characterization should have high priority in the MWFA. However, the committee also recognizes that detailed chemical analysis of such heterogeneous wastes can be very expensive and time consuming and that the nature and storage conditions of the waste present risks to operators who must obtain representative samples for analysis. The committee, therefore, recommends the following:

- The MWFA should develop simplified methods to characterize the waste, with emphasis on nondestructive examination and assay techniques.[8] According to available inventory data, emphasis should be placed on developing better methods to determine heavy metals and solvent contamination in the waste.

---

[7] The MWFA's technical needs list is reproduced in Appendix C and is discussed in the committee's findings in Chapter 4.

[8] The committee is aware that similar efforts are underway by EM to certify compliance with acceptance criteria for waste disposal at the Waste Isolation Pilot Plant. Radiography and waste drum head-space testing are included.

- The MWFA should continue to develop, demonstrate, and encourage deployment of techniques and procedures to ensure that all new waste streams are adequately characterized.
- The MWFA should strive for a balance between the risks, benefits, and cost of detailed characterization and the risks, benefits, and cost to adapt or to develop more robust treatment technologies that can handle a wide variety of waste compositions. Both characterization and technology development efforts should be pursued. Neither detailed characterization nor robust technology provides the total answer to treatment of the EM inventory.

3

Mixed Waste Regulations

This chapter summarizes the regulations that control U.S. Department of Energy (DOE) mixed wastes and briefly explains the complex regulatory situation as it currently exists. While both the committee and DOE understand that regulations will evolve, this report is necessarily framed by current regulations for characterization of the mixed waste inventory, treatment and disposal requirements, and waste form performance criteria. The U.S. Environmental Protection Agency (EPA), U.S. Nuclear Regulatory Commission (USNRC), Department of Transportation (DOT), and individual states all exert measures of control over treatment, transport, and disposal of mixed waste. As described in this chapter, the range of regulatory approaches and resulting regulations create substantial challenges for the treatment and disposal of mixed wastes.[1]

The first section of this chapter describes the treatment and disposal regulations. These regulations may directly limit DOE's options for selecting a type of waste form by imposing controls on the methods to produce the waste form or by requiring that it pass specific chemical or physical tests. The regulations may also affect the choice of waste form indirectly by imposing performance criteria on the waste form itself, on

[1] A recent announcement of proposed rulemaking describes strategies that EPA is considering to relieve mixed waste managers from some of the compliance difficulties that arise from dual regulation by EPA and USNRC (EPA, 1999a).

additional barriers intended to retain the waste, or on the entire disposal system. Performance criteria can include limitations on radionuclide release or on radiation exposure to individuals or populations.

The second section of this chapter describes waste acceptance criteria (WAC) and other requirements. The WAC are imposed by each waste disposal facility on waste it receives. Waste cannot be shipped to a facility unless it meets the facility's acceptance criteria. The acceptance criteria generally do not require specific waste forms, but they may indirectly affect DOE's options by controlling the physical and chemical characteristics of waste materials that are received.

TREATMENT AND DISPOSAL REQUIREMENTS

The two principal regulatory agencies involved in the treatment and disposal of mixed waste are the EPA and the USNRC. DOE is subject to regulations promulgated by these agencies through the Federal Facility Compliance Act of 1992 (FFCA), which requires federal facilities to comply with the same regulations as non-federal facilities (FFCA, 1992). Thus, in dealing with its mixed waste, DOE's Office of Environmental Management (EM) must comply with EPA regulations for hazardous wastes and with USNRC regulations for radioactive wastes. Further, the FFCA requires DOE to comply with applicable state regulations if they are more restrictive than federal regulations. For the transport of waste materials, DOT regulations must be observed. In addition, DOE has entered into agreements with many states, which may place added constraints on the selection of treatment and disposal options. Because of its importance for DOE mixed waste management, the FFCA is further explained in Box 1. USNRC, EPA, and DOE regulations referred to in the box are described in the subsequent sections of this chapter.

U.S. Environmental Protection Agency

EPA has developed regulations for hazardous waste management and disposal principally under authority of the Resource Conservation

Box 1
The Federal Facility Compliance Act

The Federal Facility Compliance Act of 1992 has played an important role in DOE mixed waste management programs. Two provisions of the FFCA are especially relevant to the issue of mixed waste management. First, the FFCA requires that DOE facilities comply with all federal, state, and local laws and regulations pertaining to hazardous waste. Thus, DOE facilities are subject to the hazardous waste requirements promulgated under the Resource Conservation and Recovery Act. States that have been delegated authority by EPA to manage their own hazardous waste programs have authority over DOE facilities.

The FFCA did not alter the separation of Atomic Energy Act (AEA) authorities and responsibilities between DOE and the USNRC. Radioactive waste is defined by the AEA. Thus, the AEA portion of EM's mixed waste is regulated by DOE under terms of DOE Order 5820.2A and forthcoming Order 435.1 that will supersede it. The USNRC has no jurisdiction over defense transuranic (TRU) waste and the Waste Isolation Pilot Plant, but does have jurisdiction over commercial TRU waste (one form of Greater-Than-Class-C waste that must be disposed by DOE with USNRC approval). DOE continues to define low-level waste (LLW) by exclusion [that which is not high-level waste (HLW), TRU waste, etc.] while the USNRC uses the formal classification of LLW found in 10CFR Part 61.

The second provision of the FFCA of relevance to DOE MLLW disposal programs is the requirement to develop and submit site treatment plans (STPs) for mixed waste at each facility at which the DOE stores or generates these wastes. These plans identify how the DOE will provide waste treatment for all MLLW waste streams and must include schedules for bringing new facilities into operation. The DOE identified 35 sites requiring STPs and submitted Draft STPs for nearly all of these sites in October 1995.

Since its passage, the FFCA has provided a strong incentive for the DOE to work with the states to develop management and disposal strategies for mixed HLW, LLW, and TRU waste. A number of initiatives were begun that have significantly improved the relationship between these two parties with respect to waste management.

and Recovery Act (RCRA), enacted in 1976.[2] RCRA has been amended several times, with the most significant amendments passed in 1984 as the Hazardous and Solid Waste Amendments. RCRA provides for the cradle-to-grave control of hazardous wastes by imposing management requirements on generators and transporters of hazardous waste and on owners and operators of treatment, storage, and disposal facilities. Regulations pertaining to RCRA waste disposal facilities (landfills) include such details as liner and cover designs. Figure 2 illustrates a RCRA landfill.

The RCRA hazardous waste regulations are found in Title 40 of the Code of Federal Regulations (EPA, 1996). Parts 260 to 265 describe hazardous waste management, provide EPA's lists of hazardous wastes, and set standards that must be met by hazardous waste generators and managers. EPA's land disposal restrictions are given in 40CFR268 and its permit programs in 40CFR270.

Hazardous wastes are defined in 40CFR261. Waste materials are classified as hazardous in two ways. The first way is if constituents in the waste appear on comprehensive lists including more than 700 materials, representing non-specific wastes (such as electroplating wastes or spent solvent wastes),[3] unique wastes from specific industries, and commercial chemical products including residues and spills. These are known as "listed wastes." The second way is for the waste material to exhibit hazardous characteristics of ignitability, corrosivity, reactivity,[4] or toxicity; these wastes are referred to as "characteristic wastes."

The toxicity characteristic is defined by a list of 40 materials, comprised of seven metals (including lead, chromium, cadmium, and mercury)[5], and 33 pesticides and solvents (including chlorinated solvents commonly found in industrial wastes). A waste material may be declared

---

[2] A history of EPA's regulation of mixed waste beginning in 1976 can be found on the EPA Mixed Waste Team internet home page: http//www.epa. gov/radiation/mixed-waste/.

[3] More than 80% of the EM mixed waste inventory is listed as hazardous because of these materials.

[4] Reactive materials may pose a fire or explosion hazard in the course of handling or disposal.

[5] More than 50% of the EM mixed waste inventory is listed as hazardous because it contains these metals.

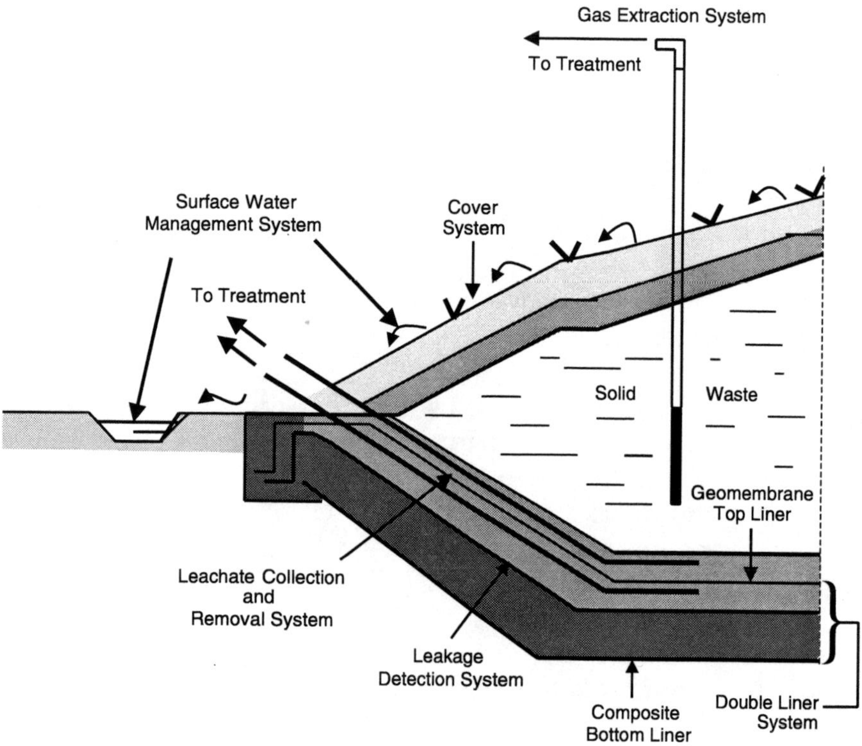

FIGURE 2 Illustration of a RCRA Landfill. SOURCE: EPA, 1999b

hazardous based upon knowledge of process chemistry or it may be tested. A common test for the toxicity characteristic as well as for many listed waste constituents is the Toxicity Characteristic Leaching Procedure (TCLP), which involves the extraction of the solid waste material by a weak acetic acid solution. The presence of characteristic wastes or listed wastes in the extract in excess of established concentration limits confirms the solid waste to be a hazardous waste. Untreated EM wastes will often include a combination of both listed and characteristic wastes.

A waste that has been determined to be hazardous is subject to the provisions of the Land Disposal Restrictions (LDRs) as described below.[6] Characteristic waste may be treated to exit the hazardous waste management system. Listed wastes may be delisted, but may still be subject to the LDRs.[7] A more thorough discussion of the TCLP in the context of waste form characterization is provided in Chapter 5.

LDRs in 40CFR268 apply to all hazardous wastes, including mixed wastes. LDRs either identify the maximum concentrations of hazardous constituents in waste that may be disposed of in a permitted hazardous waste landfill or prescribe ways that the waste may be treated to allow disposal in a permitted hazardous waste landfill. Prescriptive treatment methods are referred to as Best Demonstrated Available Technology (BDAT).[8] BDAT applies to three categories of radioactively contaminated wastes: (1) lead solids; (2) elemental mercury; and (3) hydraulic oils contaminated with mercury (Sullivan, 1997). In addition to BDAT treatment for low-level mixed waste, other BDAT treatment standards for hazardous waste will apply to mixed waste where specific waste constituents are identified.

For most waste constituents of direct concern to EM, BDAT has not been identified, but treating the waste to meet the Universal Treatment Standards (UTS) will allow disposal in a landfill. The UTS specify the maximum allowable concentration of waste constituents in the treated waste, either by direct analysis or application of the TCLP procedure, but do not specify the treatment process. In addition, EPA recognizes that some mixed wastes may not be treatable by either by

---

[6] Currently EPA is preparing a "Hazardous Waste Identification Rule" (HWIR) to amend certain portions of the RCRA regulations. The purpose of the HWIR is to exempt from hazardous waste regulations those materials currently designated as hazardous waste if they contain hazardous constituents only at concentrations that pose very low risk to humans and the environment. These wastes could then be disposed of in non-hazardous (Subtitle D) landfills or other disposal units.

[7] Regulations in 40CFR260.22 provide methods for "delisting" or removing specific wastes or treated wastes at a given facility from provisions of the hazardous waste program. In general, the petitioner must show through comprehensive testing and analysis that the specific material does not meet the criteria that caused it to be listed by the (EPA) Administrator.

[8] 40CFR268.42

BDAT or to the UTS. In such situations, EPA will allow petitions to be submitted to request a variance.

The Toxic Substances Control Act (TSCA) of 1976 (as amended) addresses materials in wide use that pose an extraordinary hazard to persons who may come in contact with them. Examples include polychlorinated biphenyls (PCBs) and asbestos. TSCA gives EPA authority to regulate the manufacture, use, distribution, and disposal of these hazardous materials. In addition, TSCA includes provisions for handling and cleaning up wastes containing PCBs. These provisions apply even if the wastes were generated before the effective date of TSCA. As discussed in Chapter 2, a small but significant amount of EM's mixed waste inventory is suspected to be contaminated with PCBs.

To put EPA's hazardous waste regulatory approach in perspective, one must recognize that its regulations apply to over 500,000 companies and individuals throughout the United States (Case, 1991). Thus, there is a strong justification for the regulations to be applicable universally and to contain straightforward numerical criteria that are relatively easy to understand and enforce. The EPA approach to assuring safety includes a definition of hazardous waste, specifies treatment standards that must be met prior to land disposal, and specifies standards for construction and operation of hazardous waste disposal sites. The specificity of the regulations does not require interpretation by DOE; only compliance is required.

In summary, the EPA regulations that apply to EM's program for managing its mixed wastes require the following:

- Systems constructed to treat mixed wastes must be permitted by EPA or the host state (40CFR264).
- Inventory reporting and other requirements of TSCA must be followed.
- Each hazardous waste storage or disposal facility must be permitted and approved (40CFR264 and 40CFR270).
- MLLW disposal must meet EPA's hazardous waste landfill regulations, including LDRs (40CFR268).
- Mixed TRU wastes are subject to EPA regulations appropriate to high-level and transuranic wastes. These require

deep geological disposal and performance assessment over a 10,000-year time period.[9]

U.S. Nuclear Regulatory Commission

The USNRC operates under the authority of the Atomic Energy Act of 1954 (AEA) and its subsequent amendments. USNRC regulations that affect management of DOE mixed wastes include the following:

- 10CFR61 Low-level Waste Disposal Regulations;[10] and
- 10CFR20 Radiation Protection Standards.

The waste disposal regulations require that releases from the disposal facility meet the radiation protection standards.

The USNRC closely regulates the waste form stability and radioactive characteristics of low-level waste materials acceptable for near-surface land disposal through a combination of prescriptive and performance-based requirements (10CFR61.55-56). According to 10-CFR61, a near-surface disposal facility is one in which radioactive waste is disposed in or within the upper 30 meters of the land's surface. Institutional control of access is required for 100 years (10CFR61.7), and within 500 years, wastes must decay to a sufficiently low level that remaining radioactivity will not pose unacceptable hazards to an intruder or the general public. To meet this latter requirement, further prescriptive regulations define three classes of waste that are deemed suitable for near-surface disposal. Classification as Class A (the least restrictive), Class B, or Class C depends on which radionuclides are present and their concentrations. The radioactive half-life is the primary discriminator (Table 2). If any long-lived nuclides are present in the waste in concentrations greater than given in Table 2, the waste is not suitable for near-surface disposal.[11] Mixed transuranic waste (MTRU) is thus

---

[9] 40CFR191 and 40CFR194

[10] Commercial disposal facilities must be licensed by the USNRC. Commercial facilities, such as Envirocare in Utah, are expected to be the final destination for about 30% of DOE's MLLW (DOE, 1997).

[11] Mining industry waste is excluded from this requirement.

TABLE 2 Allowable Concentrations of Long-Lived Radionuclides for Near-Surface Disposal

Radionuclide	Concentration, curies per m^3 (Ci/ m^3)
^{14}C	8
^{14}C in activated metal	80
^{59}Ni in activated metal	220
^{94}Nb in activated metal	0.2
^{99}Tc	3
^{129}I	0.08
	Concentration, nano-curies per gram (n Ci/g)
Alpha emitting transuranic nuclides with half-life greater than 5 years	100
^{241}Pu	3,500
^{242}Cm	20,000

SOURCE: 10CFR61.55

TABLE 3 Allowable Concentrations of Short-Lived Radionuclides for Near-Surface Disposal

Radionuclide	Class A Waste (Ci/ m^3)	Class B Waste (Ci/ m^3)	Class C Waste (Ci/ m^3)
Total of all nuclides with less that 5-year half-life	700	Note 1	Note 1
^{3}H	40	Note 1	Note 1
^{60}Co	700	Note 1	Note 1
^{63}Ni	3.5	70	700
^{63}Ni in activated metal	35	700	700
^{90}Sr	0.04	150	7000
^{137}Cs	1	44	4600

SOURCE: 10CRF61.55

Note 1: There are no limits for these radionuclides in Class B or C wastes. Practical considerations such as the effects of external radiation and internal heat generation on transportation, handling, and disposal limit the concentrations for these wastes. These wastes shall be Class B unless the concentrations of other nuclides in Table 2 determine the waste to be Class C independent of these nuclides.

generally excluded from near-surface land disposal by USNRC regulations, and requires deep geological isolation. Disposal of MTRU in the Waste Isolation Pilot Plant (WIPP) repository is described at the end of this chapter. In the absence of long-lived radionuclides, near-surface disposal of substantial concentrations of short-lived nuclides is allowed according to the provisions in Tables 2 and 3. [12]

The USNRC historically has used performance-based standards in addition to prescriptive requirements. The original USNRC standards were developed for occupational exposure and were based on the probability of one premature death per year among 10,000 exposed individuals (Gershey, et al., 1990). Rather than establish designs for a disposal facility or establish strict numerical standards for each radionuclide, USNRC regulations establish performance objectives (10CFR61, Subpart C) that define regulatory limits (accepted health-based standards) for the radiation exposure that a person in the vicinity of a disposal site may receive. A disposal facility must be able to perform its function of retaining radionuclides well enough to meet these objectives. A technical analysis is required to demonstrate that a specific system can meet the performance objectives.

The USNRC has responded to the need for a technical analysis through the development of performance assessment (PA) methodology (USNRC, 1997). PA can be used to demonstrate compliance with USNRC's performance objectives for radiological protection of the general public established under 10CFR61.41. The DOE has also adopted the use of PA methods for determining compliance with its radioactive waste regulations. Because the PA is essential for demonstrating the long-term safety of a disposal system, the methodology is described in Chapter 6 of this report.

---

[12] When waste contains several radionuclides, but none exceeds the limits in Table 2 or 3, the fractions of the limit for each nuclide are summed. If the sum exceeds 1.0, the waste is deemed unsuitable for near surface disposal. For example, waste with a concentration of 2 curies of ^{99}Tc per cubic meter and 70 nanocuries of transuranic nuclides per gram would have a sum of fractions of 1.36 (i.e., 2/3 + 70/100). It would be prohibited from near-surface disposal. Requirements for wastes containing long-lived nuclides having a sum of fractions greater than 0.1 include structural stability and additional disposal conditions to guard against inadvertent intrusion.

For making a PA, USNRC regulations place emphasis on siting criteria, and, in particular, recognize the importance of ground water flow as a contaminant transport mechanism (10CFR61, Subpart D). USNRC regulations state that "site characteristics should be considered in terms of the indefinite future and evaluated for at least a 500-year time frame" (10CFR61.7). Recent USNRC documents suggest that a 10,000-year horizon may be an appropriate end point for PA studies covering the disposal system for low-level mixed waste (USNRC, 1997). As will be discussed in Chapters 5 and 6, there are no available tests that can reasonably predict the very long term performance of waste forms. As a result, performance assessments often take little credit for retention of long-lived radionuclides by the waste form.

U.S. Department Of Energy

In compliance with the FFCA, DOE regulates cleanup and waste disposal at its sites through implementation of internal orders based on its authority under the AEA. These orders include the following:

- DOE Order 5400, General Environmental Protection (DOE, 1998a); and
- DOE Order 5820, Radioactive Waste Management.

DOE is required to manage its mixed wastes in a manner that assures the health and safety of site operating personnel and the public, and protection of the environment. Waste treatment must minimize the generation of secondary wastes and comply with all applicable federal, state, and local environmental, health, and safety laws and regulations.

DOE Order 5400.5 sets radiation exposure standards for the public and on-site personnel. These standards permit a total of 100 millirem per year (mrem/yr) exposure to an individual member of the public. Of this total, only 10 mrem/yr can result from inhalation of airborne particles, and exposure from ingesting radionuclides in drinking water is limited to 4 mrem/yr[13] Any disposed radioactive or mixed waste

[13] The remainder would come mainly from sources outside the body (i.e., external doses).

must not contribute to the exposure of the public in excess of these amounts. Permissible levels of radioactivity would, therefore, depend on the ability of the disposal system to retain the radionuclides. Additionally, all exposures must be limited to "as low as reasonably achievable" levels.

DOE Order 5820.2A deals with radioactive waste management, including high-level waste, TRU waste, and LLW, and contains a number of provisions that meet and extend the regulatory criteria established by the USNRC. The order, promulgated in 1988, requires that the concept of waste minimization, including waste segregation, be applied in process design and operation. The order also requires a PA for all disposal facilities. There are two objectives that must be addressed in PA models (DOE, 1996c).

The first objective is to assure that the disposed waste will not cause an exposure (effective dose equivalent) greater than 25 mrem/yr to any member of the public. DOE has interpreted this to include all pathways of possible exposure during the period of 1,000 years following closure of the disposal facility. To increase confidence in the outcome of the modeling, analysis beyond 1,000 years (but not exceeding 10,000 years) may be done (DOE, 1996c).

The second applicable performance objective addresses inadvertent intrusion after the institutional control period (100 years), requiring that the effective dose equivalent received by an individual who might inadvertently intrude into the facility not exceed 100 mrem/yr for continuous exposure or 500 mrem for a single acute exposure. PA modeling of this scenario beyond 1,000 years post-closure is not recommended.

WASTE ACCEPTANCE CRITERIA AND OTHER REQUIREMENTS

In addition to the waste treatment and disposal regulations that were discussed in the previous section, controls may be imposed at waste treatment or disposal sites through the mechanism of WAC. All DOE facilities are required to have WAC for waste received by that facility for treatment or disposal (DOE Order 5820.2A). As indicated by the name, WAC describe requirements that are placed on received waste in order that it can be accepted at a facility. Wastes delivered to any disposal site are required to meet LDR and any conditions imposed by the WAC or the facility permitting process. The criteria may address shipping requirements; containerization; physical, nuclear, and chemical properties, including gas generation; and waste characterization. Each facility develops specific WAC to ensure that it operates safely and will meet all applicable regulations during operation and after closure. For non-DOE facilities, such criteria may be incorporated in the facility permit or specified separately. In the case of treatment facilities, the WAC do not establish requirements on the waste form for disposal but rather on the waste as received in anticipation of further treatment prior to disposal. WAC are site specific, and may place additional constraints on the selection of waste forms for disposal.

Except for the Envirocare of Utah facility, no commercial options are available for the land disposal of DOE mixed waste. At Envirocare, the WAC place low limits on the radioactivity of waste that can be received, thereby restricting the use of this site. Although the DOE sites at Hanford and the Nevada Test Site can dispose of mixed wastes generated on-site, they are not currently available for mixed waste generated at other sites. The WIPP project may accept waste meeting the definition of TRU and MTRU waste, but it may not receive MLLW. Since disposal sites are not available for the majority of MLLW, DOE is faced with having to design treatment systems and subsequent waste forms without knowledge of potential constraints that may be imposed in the future. Waste treatment and temporary storage of the waste forms until a disposal facility is available may be necessary to meet EM's cleanup schedule. However, this expedient is more uncertain and expensive compared with treatment followed immediately by final disposal, and imposes an additional set of regulations applicable to temporary storage facilities.

Shipping regulations imposed by DOT may add constraints with regard to packaging of waste material for transport if wastes must be shipped to an off-site location for treatment or disposal. A description of the effect of WAC and transportation requirements on TRU waste forms for the WIPP facility is given in Box 2.

Box 2
The Role of Waste Forms at WIPP:
A Special Case for Mixed TRU Waste

The Waste Isolation Pilot Plant (WIPP) is an underground facility designed for the disposal of DOE TRU waste. The repository is located in a semi-arid region of southeastern New Mexico. WIPP is constructed in bedded salt at a depth of approximately 658 m (2,160 ft) (see Figure 3). A description of the WIPP facility, the site hydrogeology, the waste characteristics planned for the facility, and the regulatory issues that must be met by the facility was published by the National Research Council's Committee on WIPP (NRC, 1996b).

WIPP is expected to be the disposal facility for nearly all of DOE TRU and mixed TRU waste.[†] The design capacity of WIPP is 168,500 m^3 of contact-handled waste and 7,000 m^3 of remote-handled waste. The principal radioactive constituents of the waste will be uranium and isotopes of the transuranic elements neptunium, plutonium, and americium with half-lives greater than 20 years and concentrations greater than 100 nCi/g (DOE Order 5820.2A).

In 1992, the WIPP Land Withdrawal Act (Public Law 102-579) transferred control of land at the WIPP site from the Department of the Interior to the DOE. A major provision of the Act requires DOE to show that the repository will comply with applicable federal regulations. The principal regulations that WIPP must satisfy are the EPA general regulations for spent fuel and TRU wastes in 40CFR191, specific regulations developed for WIPP in 40CFR194, and RCRA 40CFR Chapter I provisions pertaining to management and disposal of hazardous wastes. WIPP is also subject to DOE regulations in DOE Order 5820.2A.

Congress passed amendments to the Land Withdrawal Act (Public Law 104-201) that exempted WIPP from treatment standards and LDR requirements. Thus, wastes do not require any form of stabilization for disposal in the WIPP. All waste will be placed in 55-gallon steel drums or in standard waste boxes, which will be placed in DOT-approved TRUPACT II canisters for transportation to WIPP. The waste will be disposed of in the same drums or waste boxes. Remotely handled TRU waste will be placed in approximately 1 m^3 steel canisters. The waste will be placed in mined-out disposal rooms that are approximately 4 m high,

10 m wide and 100 m long. It is anticipated that salt creep will completely close the rooms and compact the waste within 100 years.

Gas generation resulting from the decay of organic matter in the waste is a concern because it might lead to build-up of high pressures in the repository. This could compromise containment of the wastes in the event of a drilling intrusion into the disposal region. Magnesium oxide (MgO) will be emplaced with the waste to limit the effects of gas generation by forming a precipitate with CO_2.

The National Research Council's Committee on the Waste Isolation Pilot Plant reviewed the status of WIPP and its suitability as a repository for TRU waste (NRC, 1996b). Among other issues, the WIPP committee considered the issue of waste form modification and whether significant benefits could be achieved by additional waste treatment or stabilization. The WIPP committee noted that:

- Extra costs would be incurred to treat the waste.
- There would be more risk to treatment plant personnel than if untreated waste were being disposed.
- A waste treatment system would take significant time to implement.
- Treatment options would involve more complicated regulatory requirements.

This led the WIPP committee to agree with DOE's conclusion that additional waste treatment or stabilization is not desirable in the special case of the WIPP.

---

† The WIPP received its first shipment of TRU waste on March 26, 1999, as this report was being prepared for publication. To receive mixed TRU waste, the WIPP lacks only a hazardous waste permit, which must be issued by the New Mexico Environment Department.

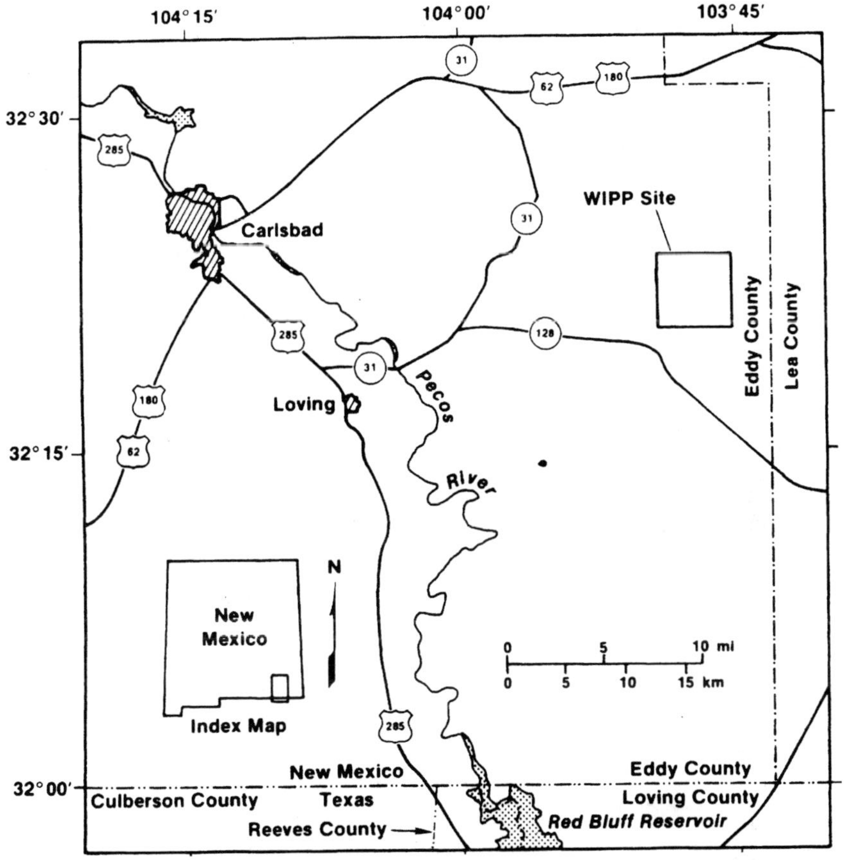

FIGURE 3 Location of the Waste Isolation Pilot Plant site. Inset map shows approximate location of map area in the state of New Mexico.
SOURCE: Modified slightly from Sewards, et al. (1991, Figure III-1).

COMMENTS AND RECOMMENDATION

The state of development of technologies for treating mixed waste and converting it to stable waste forms must be assessed in the context of the regulations that control these wastes. The committee found that the absence of an agreed-upon set of guidelines for acceptance of waste forms at future disposal facilities and for performance of waste forms in the disposal environment introduces significant uncertainties into EM's technology development programs.

The unavailability of existing disposal facilities that have USNRC licenses and EPA or state permits to accept mixed waste places uncertainties on the selection of treatment processes and waste forms. Currently, EM must make assumptions about the acceptability of its intended waste forms at expected disposal sites in order to complete the design of treatment systems. Treatment system modification or additional treatment of the waste may be required if the expected site does not become available, or if the waste form does not meet unanticipated WAC at a new site.

Although the regulatory goals of USNRC and EPA are the same, protection of health and the environment, the committee noted that fundamentally different approaches are used by the two agencies to achieve this goal. EPA regulations are applied to a large range of waste streams. Therefore, EPA has adopted prescriptive regulations with options for exception. In contrast, USNRC and DOE regulations apply to a relatively small number of waste streams, allowing individualized licensing and permitting to conform to performance standards. These differences result in apparent inconsistencies between the respective regulations, which make management of mixed waste difficult.

The USNRC and DOE require that the waste form, packaging, and emplacement ensure stability and performance predictability. The essential measure of disposal acceptability is the performance assessment that is used to estimate the dispersion of the radioactive species from the disposed waste over a long time scale, and to compare this estimate against some standard of acceptable off-site effects. In addition, the disposal site must be retained in perpetual custody by the state or DOE. [14]

The EPA regulatory approach is prescriptive: define the hazardous waste by listing or by its hazardous characteristics as measured by the TCLP, set waste form requirements for resistance to leaching, and require that the disposal facility include features such as a cover, a dual liner, and a leachate collection system. The design requirements for the disposal facility and its contents are such that virtually no leaching should occur in the 30-year performance period.

---

[14] Section 11.e(1) in the Atomic Energy Act of 1954.

Some further specific examples of these differences are the following:

- *Disposal facility design.* EPA has developed design criteria for hazardous waste disposal facilities that apply regardless of the geologic and climatic setting. USNRC does not identify or recommend any particular design, but instead establishes performance objectives for the entire disposal system. The performance objectives include consideration of site geology and climate. Joint USNRC-EPA guidance for conceptual design of MLLW disposal facilities has been written (EPA, 1987).
- *Disposal facility approval.* EPA's permitting procedure for establishing a disposal facility includes site characteristics but not waste characteristics.[15] The USNRC requires a performance assessment to demonstrate that a disposal facility will meet its performance objectives. The performance assessment must consider both site and waste characteristics.
- *Waste concentration limits.* For near-surface disposal of any waste, EPA sets maximum concentration limits for hazardous materials in the waste (either directly or through application of a BDAT treatment) or requires that the waste form pass the TCLP. The USNRC uses a waste classification system that allows near-surface disposal of waste containing only very low levels of long-lived radionuclides but relatively high concentrations of short-lived radionuclides.
- *Period of performance.* EPA requires a disposal facility to contain all waste constituents during a minimum 30-year monitoring period after closure.[16] The USNRC requires preparation of a formal performance assessment to demonstrate that the site will meet quantitative performance objectives. The

---

[15] For facility operation, waste characteristics must be known, for example, to ensure that a given waste will not react with other wastes, that it is compatible with the containment system, and that it has been treated to LDR levels prior to disposal.

[16] At the end of the 30-year period, the EPA administrator may extend the post closure period if deemed necessary for protection of human health and the environment.

USNRC has recommended that the period of performance considered in the PA be at least 1,000 years and may be extended to 10,000 years or more.

- *Waste form performance.* Neither the DOE, EPA, nor the USNRC provides guidance for waste form performance in the repository environment. As noted earlier, EPA requires only that waste forms pass the TCLP. Performance assessments required by USNRC and the DOE generally give little credit for waste containment to the waste form. This lends conservatism to the calculations of disposal facility performance as will be discussed in Chapter 6.

The committee recommends that because of the lack of available disposal sites for MLLW and the difficulties in establishing new sites, EM should work with EPA and the USNRC to agree on clear guidelines that describe acceptable waste forms for disposal of mixed waste in future, near-surface disposal facilities. This should be done as soon as possible to reduce the risk of EM deploying technologies that are later judged inadequate because of unanticipated regulatory requirements.

4

Waste Treatment and Stabilization

This chapter deals with the central feature of the committee's task to review and evaluate the state of development of the final forms of treated wastes as they arise from current and emerging treatment technologies. The first section of the chapter presents the Mixed Waste Focus Area's (MWFA's) approach to identifying appropriate technologies for treating and stabilizing mixed waste, namely the division of the U.S. Department of Energy (DOE) Office of Environmental Management's (EM's) mixed waste inventory into five groups so that the wastes in each group have similar treatment requirements. Plans for obtaining the technical services of private contractors (privatization) to treat portions of the inventory are described. The second section gives an overview of the current and emerging treatment technologies for each waste group. Waste forms that result from the treatment processes or that can be made from the treated waste are described in the third section. Needs in technology identified by the MWFA that are directly related to waste form development are discussed in the fourth section. The committee's findings and recommendations are presented in the final section.

The MWFA methodology to identify and provide mixed waste treatment technologies in support of EM's cleanup goals were presented in two documents, both entitled "Mixed Waste Focus Area Technical Baseline Report" (DOE, 1996a and DOE, 1997a). The Technical Baseline Reports provide information about the state of development of technologies for the treatment of mixed waste and production of acceptable waste forms. Most of the information in this chapter was taken from the Technical Baseline Reports and presentations made to the committee by MWFA representatives. All numerical data were taken from the 1997 Baseline Report.

TREATMENT GROUPS

The present inventory of mixed wastes under EM's responsibility is about 167,000 cubic meters (m^3). Approximately two-thirds of the waste is mixed low-level waste (MLLW) and the remainder is mixed transuranic (MTRU) waste. The inventory is expected to grow to about 250,000 m^3 during the next few years, and that inventory will be further increased by waste resulting from EM's site remediation activities (Kolts, 1996). While these additions will increase the total inventory, the current inventory is believed to be representative of all the various types of mixed wastes that must be treated and stabilized. EM's mixed waste inventory was described in Chapter 2.

The MWFA has categorized the current inventory of EM's mixed waste into five groups, based on waste characteristics that require similar handling, treatment, and associated activities. This allows the assignment of generalized treatment technologies (for example, waste water treatments) to an entire category of wastes rather than to individual waste streams. The percent of the total inventory that each group comprises, by volume, is listed in the first column of Table 4. The next three columns summarize plans for assigning responsibility for waste treatment.

Obtaining waste treatment services through competitive procurements from private contractors (privatization) is an important part of DOE's strategy for treating mixed wastes. According to Table 4, there are firm plans to treat only about 7% of the inventory in DOE facilities. Present plans are for private contractors to design, build, and operate facilities for treatment of 38% of the inventory. This leaves 55% of the inventory without a current treatment plan. Of this unassigned inventory, MWFA expects that treatment contracts for 40% will be awarded to commercial contractors. All together this will result in about 60% of the

TABLE 4 Treatment Groups for EM Mixed Waste

Waste Group	Percent of EM Inventory	Percent to be Treated in an Existing DOE Facility	Percent to be Treated via Private Contract	Percent Unassigned
Waste water	4	2		2
Combustible organics	1	1		
Inorganic, homogeneous solids and soils	47	2	7	38
Debris	46	2	31	13
Unique	2			2
Totals	100	7	38	55

SOURCE: DOE (1997a). All percentages are by volume.

total waste inventory being treated through privatization.[1] The Advanced Mixed Waste Treatment Project being constructed at the Idaho National Engineering and Environmental Laboratory (INEEL) by the private contractor British Nuclear Fuels, Ltd., exemplifies a major privatization initiative, as described in Box 3.

Box 3
Advanced Mixed Waste Treatment Project at INEEL

INEEL is currently faced with the task of treating and disposing of 65,000 m^3 of mixed wastes, under an agreement between the State of Idaho and DOE. Due to the technical and regulatory complexities associated with this waste, the DOE has contracted with a consortium of private companies to build and operate a treatment facility known as the Advanced Mixed Waste Treatment Project (AMWTP). The principal objective of the AMWTP is to produce

[1] The entire 38% in column 4 and 40% of the total in column 5, Table 4.

stabilized transuranic (TRU) waste and MTRU waste that meets the requirements for being transported from INEEL to the Waste Isolation Pilot Plant (WIPP) and probably at least one other disposal facility. The WIPP is described in Chapter 3, Box 2. Disposal plans for non-TRU waste from INEEL are yet to be determined.

To meet its objective, the AMWTP is intended to achieve a 65% volume reduction of waste at INEEL, and to provide stabilized and packaged waste that complies with WIPP waste acceptance criteria, Department of Transportation TRUPACT II, and the U.S. Environmental Protection Agency (EPA) Land Disposal Restriction requirements. The AMWTP will be designed and constructed by a consortium of companies led by British Nuclear Fuels, Ltd. (BNFL). The initial value of the contract is for $876 million to design, permit, construct, retrieve the waste, and operate the facility to treat the 65,000 m^3 of INEEL material. The estimated cost of the treatment facility alone is $270 million for design and construction. The facility is expected to open in 2003. In the intervening time, INEEL will package and ship 3,100 m^3 of waste that already complies with waste characterization, transportation, and disposal regulations to the WIPP.

The waste to be processed at the AMWTP consists of a wide variety of materials including contaminated soils, sludges and slurries, chemicals, and laboratory waste. TRU or MTRU waste that can be disposed in the WIPP comprises about two-thirds of the waste volume. Most of this waste is in boxes that cannot be shipped to the WIPP in the required TRUPACT II containers, and will have to be repackaged. In addition, much of this waste requires further characterization in order to meet the WIPP waste acceptance criteria. In its present form, the remaining one-third of the waste does not qualify as TRU waste because its concentration of transuranic nuclides is below 100 nanocuries per gram.

Figure 4 shows the proposed treatment and stabilization process for the AMWTP. Waste received at the AMWTP will be characterized to determine how it is to be processed, then separated into one of two treatment trains depending on whether it is contained in drums or boxes. The initial processing will consist of opening the containers and separating the contents by size. Approximately 25% of the waste is believed to be sufficiently homogeneous that it does not require this step, and will be subjected to volume reduction through supercompaction. Once sorted, the waste will be treated or stabilized directly. Wastes containing high concentrations of organics or hazardous compounds such as solvents and PCBs will be incinerated. The ash will be grouted to meet EPA requirements. Debris and lead-containing wastes will be macroencapsulated, although the method of macroencapsulation has not yet been chosen. Most of the waste to be treated by the AMWTP will be classified as contact handled waste providing a total exposure of less than 200 millirem per hour.

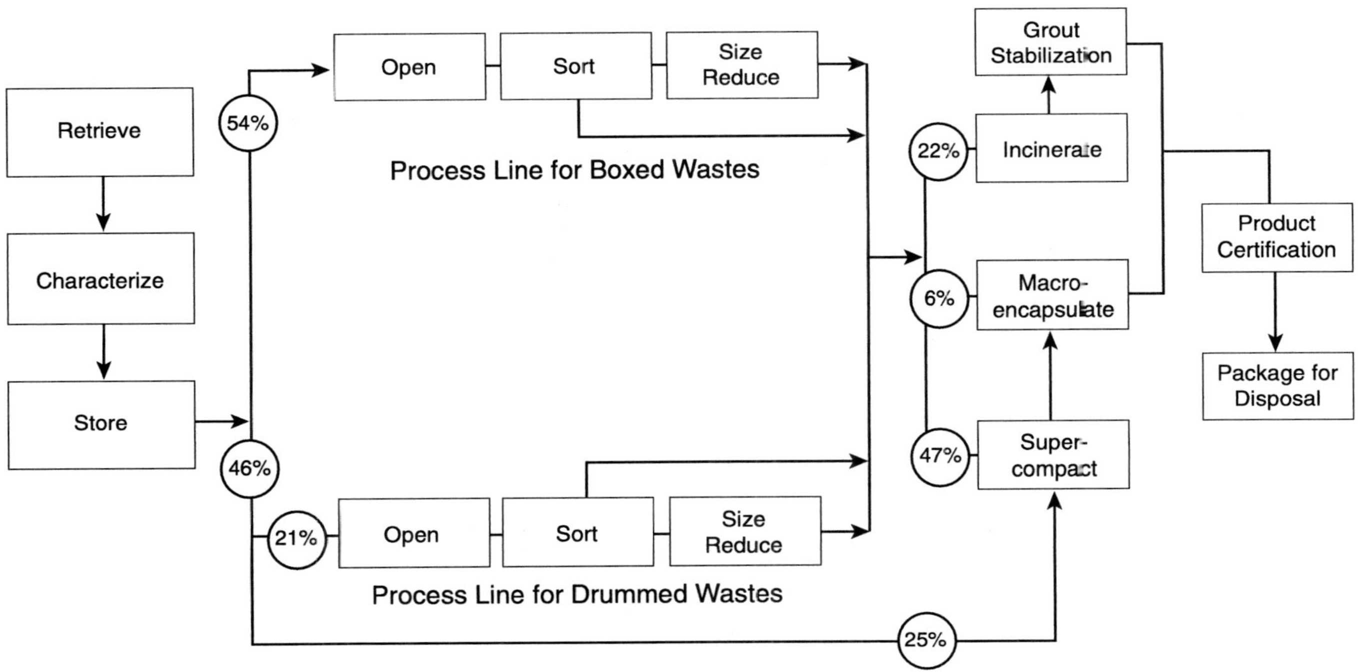

FIGURE 4 Flow diagram for waste treatment and stabilization for the Advanced Mixed Waste Treatment Process.

TREATMENT TECHNOLOGIES

The MWFA has taken current site treatment plans[2] and assembled block treatment diagrams that illustrate the treatment possibilities for each treatment group. Figure 5 provides an example block diagram of the three treatment options that MWFA has identified for the waste water treatment group. In addition to the block diagrams, basic flowsheets and processing technologies have been proposed for the treatment of about 90% of the waste inventory. Flowsheets for the remainder are yet to be proposed. Much of this remainder is in the debris group, a major fraction of which is expected to be treated through privatization. Most of the treatments described in the Baseline Report (DOE, 1996a, 1997a) are derived from processes developed for sanitary wastes or for hazardous wastes under the Resource Conservation and Recovery Act (RCRA) or Comprehensive Environmental Response, Compensation, and Liability Act (CERCLA).[3] A large number of treatment processes and a considerable number of possible waste forms resulting from these processes are applicable to mixed wastes. Table 5 summarizes the treatment and waste form options that MWFA has identified for its five mixed waste groups.

An important part of the treatment strategy is to remove, stabilize, or destroy hazardous components of each waste stream, especially removal of heavy metals and destruction of organic materials (DOE, 1996a, 1997a). The treatment strategy must lead to a waste form that satisfies the requirements discussed in Chapter 3. Volume reduction is also desired for most wastes. In an initial or pre-treatment step, solid materials may undergo size reduction and aqueous streams may be filtered to remove solids. After pre-treatment, the waste can be converted to its final form directly with such techniques as grouting and polymer encapsulation, to be described in the following section. Alternatively, the waste can be treated by thermal, physical, chemical, or biological

---

[2] Site treatment plans are required by the Federal Facility Compliance Act (FFCA) to be prepared by each DOE site. Each plan lists the wastes at the site and the treatment or disposal methods planned to bring the site into compliance with regulations. The FFCA and regulations that apply to mixed waste are discussed in Chapter 3.

[3] The RCRA and CERCLA are discussed in Chapter 3.

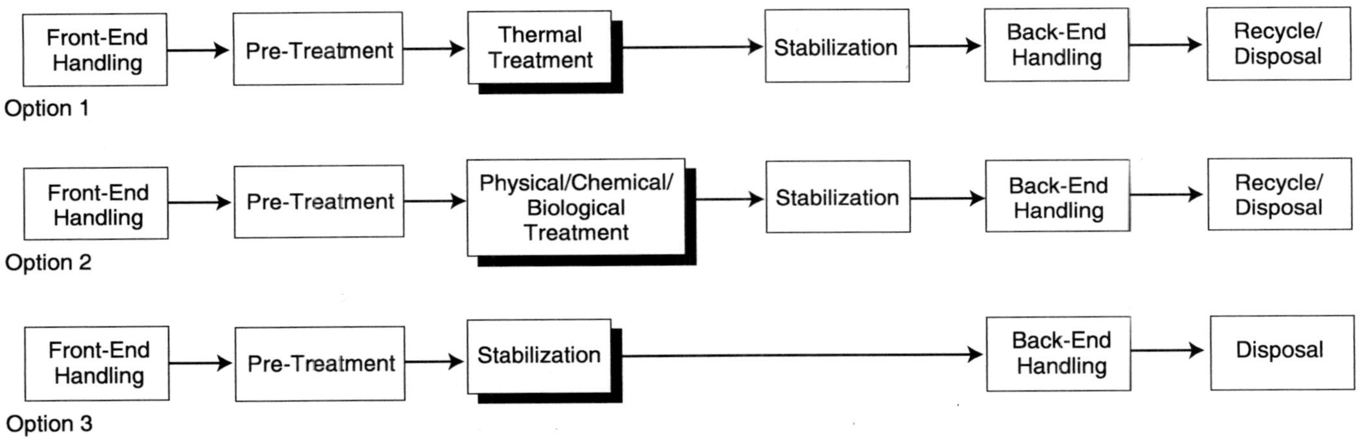

FIGURE 5 Treatment options for the waste water group.

TABLE 5 Summary of Treatment and Waste Form Options for MWFA Waste Groups

Waste Group	Hazardous Characteristics	Typical Hazardous Components	Treatment Goal	Treatment Method	Available Waste Forms
Waste water (<1% organic)	Corrosive, toxic	Cr, Pb, Cd, Hg	Volume reduction organic removal	Incineration, traditional water treatments: (reverse osmosis, neutralization, precipitation) none	Grout, polymer, glass, Hg-amalgamation
Combustible organics	Ignitable, corrosive, toxic	Halogenated; non-halogenated solvents Cr, Cd, Pb, Hg, PCBs	Destroy organics volume reduction	Incineration thermal oxidation	Grout, polymer, glass, Hg-amalgamation
Inorganic, homogeneous solids and soils (<60 mm particles)	Toxic	Electroplating waste, solvents, Pb, Cr, Cd	Volume reduction meet disposal requirements	Incineration thermal oxidation none	Grout, polymer, glass, sulfur cement
Debris (>60 mm pieces)	Toxic	Pb, solvents	Volume reduction meet disposal requirements	Thermal: incineration, melting, plasma-torch non-thermal: cutting, sorting, segregating, compaction	Grout, polymer, glass, Hg-amalgamation, direct disposal of object or compacted material
Unique	Ignitable, reactive, toxic	Reactive metals, compressed gases, explosives	Hazard reduction	Specific treatments for individual wastes or waste steams	Grout, polymer, glass, Hg-amalgamation, direct disposal

SOURCE: DOE (1997a).

methods to remove or concentrate hazardous and radioactive constituents and to put them into a form that meets regulations. These treatment processes may produce secondary wastes that themselves must be treated. Thermal treatment, for example, produces an off-gas stream that may contain metals, organic vapors, acids, or radioactively contaminated particles. These must be trapped, stabilized, and eventually disposed.

Assurance that a proposed treatment process will operate safely is required under DOE Order 5820. This order requires documentation of the waste stream analyses, treatment options, rational for process selection, and a thorough safety analysis (Safety Analysis Report). Documentation of operating and maintenance procedures, operator training, and emergency response plans is also required. The order applies to all DOE sites, their contractors, and subcontractors.

Baseline Treatment Technologies

The treatment processes described below are general descriptions taken from experience with sanitary and hazardous (i.e., RCRA) wastes. In MWFA planning, these processes are presented as available baseline technologies that may be applied to mixed waste (DOE, 1997b).

Waste water

The waste water treatment group shown in Table 5 includes aqueous liquids and slurries containing less than 1% organic material. Waste water comprises about 4% of the EM waste inventory. Volume reduction and organic removal are the major objectives for treatment of waste water. About 40% of aqueous wastes are proposed to be pre-treated and stabilized directly with cements, polymers, and, in the case of some transuranic (TRU) wastes, with glass.

Physical, chemical, and biological treatment may apply to about 48% of the waste waters. In this series of processes, organics are removed by standard water treatment techniques, such as reverse osmosis, chemical or biochemical oxidation, steam stripping, and activated carbon. Volume reduction and metals separation occur by neutralization, precipitation, filtration, evaporation, ion exchange, and other waste water treatment methods. The resulting concentrates are proposed to be grouted or incorporated into polymer matrices. Mercury is to be separated and amalgamated.

Pre-treatment and incineration are proposed for the remaining 12% of the waste water. In this process, organic compounds are destroyed, leaving an ash residue containing inorganic salts, metals, and radionuclides. Secondary wastes from this process include filtered solids, volatilized particulates and mercury, activated carbon, and ion exchange resins. All of these secondary wastes must be treated and disposed. Ash residues can be stabilized with grout, polymers or glass, and mercury can be amalgamated.

Combustible Organics

This treatment group comprises only about 1% of the EM inventory and includes organic liquids and sludges. The primary objectives for their treatment are to destroy the organic materials and reduce volume. About 93% of these wastes can be destroyed by thermal oxidation, such as incineration or an equivalent technology, and about 4% of the wastes can be oxidized by non-thermal methods. Ash residues from these wastes can be stabilized with grout, polymer, or glass. Mercury can be amalgamated.

Inorganic, Homogeneous Solids and Soils

These solid wastes, about 47% of total EM inventory, are generally soil, process sludges and particulates. A large portion of this waste group is inorganic sludge. Much of this sludge has been neutralized and partially stabilized. Contaminated soils comprise about 10% of this group, or about 5% of the EM inventory. The treatment objectives are to produce a material suitable for disposal at approved sites, and secondarily to reduce the final disposal volume. The majority of the homogeneous solids will be stabilized, with or without pre-treatment, with grout, polymer, glass, or sulfur cement. About 4% of the waste can be treated thermally through incineration, thermal desorption, or vitrification. Secondary wastes from off-gas treatment will also require treatment. The wastes resulting from thermal treatments can be stabilized with grout, polymer, or glass.

Debris

Debris is defined under RCRA as solid material greater than 60 mm in particle size that is intended for disposal. Debris can include manufactured objects, plant or animal matter, or natural geologic material. The debris treatment group represents about 46% of EM's mixed waste. Because of its large quantity and its heterogeneity, the group presents a considerable treatment challenge.

For the majority of the materials in this group, the proposed treatment method is undetermined, but it will be selected to meet shipping and disposal requirements. Potential treatment methods are well established and include sorting and segregation, size reduction, removal of hazardous components by washing or mechanical means, and grouting of liquids. Direct stabilization of debris by macroencapsulation or grouting is also possible. Mercury can be removed by desorption and amalgamated. According to present plans about 13% of this waste may be treated thermally.

Unique Wastes

Unique wastes are those that do not fit into the other treatment groups. Waste streams comprising this group are generally small, and overall, this group constitutes only about 2% of EM's inventory. However, treatment of these wastes presents a considerable challenge in that treatment processes may have to be tailored for each waste stream. The primary objective in treating these types of wastes is detoxification in preparation for disposal.

Lab packs (mixed waste from laboratory operations) are expected to be oxidized thermally or chemically and may be also be treated by chemical precipitation. Waste products from this processing can be stabilized with grout or polymer. Mercury can be amalgamated. Metals such as lead, cadmium, or beryllium can be macroencapsulated after their surfaces have been cleaned to remove easily mobilized contamination. Reactive metals can be deactivated and residues can be stabilized.

Treatment processes for common batteries involve surface decontamination, liquid and solid separation, neutralization, and stabilization. Explosives and propellants will be incinerated or otherwise thermally oxidized or deactivated chemically. Compressed gases and aerosols will be incinerated, chemically reduced, or oxidized.

New Treatment Technologies

During the committee's review period, a primary goal of the MWFA was to demonstrate three technologies that had the potential to treat 90% of EM's mixed waste inventory. A key requirement was that the technologies produced a waste form that met all RCRA stability and leaching criteria and would not be significantly degraded by radiation damage. The technologies of interest were the following:

- Joule–heated melting;
- plasma–torch melting; and
- macroencapsulation.

According to presentations by the MWFA to the committee, tests of these technologies confirmed their potential for treating most of the mixed waste inventory. However, the tests also showed that actual application of the technologies to treat specific waste streams would require that each of the technologies be demonstrated, operating parameters be identified and optimized, and each design be adapted for the waste stream. The extremely broad physical, chemical, and radiological properties of EM's mixed waste may preclude the use of generically designed treatment processes. Waste forms that result from use of these technologies are described, along with other available waste forms, in the next section.

In addition to the above broadly targeted technologies, three other technologies were reported by the MWFA to be in an advanced stage of development and demonstration. These include molten metal processes, non-thermal treatment processes, and improved methods to stabilize mercury.

Molten Metals

Metallic pools with dissolved reagents (e.g., oxygen) and often with liquid slags, have been used to process mixed wastes (Evans, et al., 1997; EPRI, 1997a; EPRI, 1997b). The high temperature and highly reactive environment completely destroys organic waste components, and the metallic constituents partition between the slag and the bulk metal. Metals such as iron, copper, and nickel have been used as metallic baths and commercial scale mixed waste processing has been demonstrated. The high temperature of the process requires close attention to the off-gas streams, depending on the composition of the feed to the process. The aqueous corrosion behavior of product metals, while complex, may be extrapolated with modest confidence, especially if the metals are simple, that is, not complex alloys. The robust nature of the metals and the ceramic slags points to relatively reliable extrapolation of confinement loss and release rates. The process was demonstrated at Oak Ridge. Several demonstrations on non-radioactive wastes have also been reported.

Non-Thermal Treatment Processes

DOE recently completed an evaluation of alternative nonflame technologies for destruction of organic waste and organic constituents in aqueous waste (Schwinkendorf, 1997). Many of the technologies (e.g., supercritical water oxidation) were only applicable to aqueous wastes. Some of the alternative technologies, including electrochemical or biochemical decomposition, may have niche applications to streams for which incineration or other high temperature processes are inappropriate. The evaluation recommended further work on the more aggressive technologies: steam reforming (pyrolysis at 300 °C to 1200 °C in the presence of steam), direct chemical oxidation using peroxidisulfate, and a co-catalyzed wet oxidation process using acidic ferric chloride. If these processes were to become commercially attractive, the waste products resulting from these processes would still require immobilization.

Mercury Stabilization

Treatment of mercury-contaminated waste and disposal of mercury are among EM's highest priority mixed waste problems. At least some mercury is found in all five MWFA treatment groups. As discussed in Chapter 3, the U.S. Environmental Protection Agency (EPA) prescribes treatments, referred to as Best Available Demonstrated Technology (BDAT), for mercury-contaminated wastes. Wastes that contain mercury at contamination levels less than 260 parts per million (ppm) require stabilization so that mercury released from the resulting waste form by the Toxicity Characteristic Leaching Procedure (TCLP) is below 0.2 ppm. For solids contaminated with more than 260 ppm total mercury, EPA requires that the mercury be thermally separated from the waste and recovered (retorted). This is based on the premise that the mercury would be recyclable. Since the mercury recovered from mixed waste is expected to remain radioactively contaminated, it cannot be recycled. The MFWA is working with the EPA to develop appropriate treatment standards to replace the requirement for retorting.[4]

Near the end of the committee's review period the MWFA was evaluating two promising methods of providing waste forms for elemental mercury (DOE, 1997b). The first, amalgamation, was being demonstrated on five elemental mercury wastes from four DOE sites by two private subcontractors. The second method, stabilization of elemental mercury with sulfur polymer cement, was being demonstrated at Brookhaven National Laboratory. The development of direct stabilization technologies to provide waste forms for disposing of elemental mercury recovered from mixed wastes is a priority item in the MWFA technology development needs list described later in this chapter and reproduced in Appendix C.

AVAILABLE WASTE FORMS

A wide variety of waste forms is available for stabilizing the products of the treatment processes described in the preceding section.

---

[4] Approaches to disposing of mercury are widely varied. Swedish law, for example, requires deep geological disposal of mercury (SEPA, 1998).

Selection of the most appropriate waste form for a waste stream at a DOE site is a key step in that site's overall waste management strategy. In identifying waste forms for mixed waste, the MWFA has used the considerable experience in both the private sector and DOE. The literature contains many examples of waste forms that have been developed outside the MWFA for sanitary, hazardous, and low- and high-level radioactive waste that are nevertheless applicable to mixed waste (Perret, 1998; Gilliam and Wiles, 1996; Lutze and Ewing, 1988; Donald, et al, 1997). Waste forms in MWFA's repertoire of available baseline technologies will be described in this section. The basic classes of waste forms include:

- Grout
- Glass
- Polymers
- Crystalline ceramics
- Vitreous ceramics
- Compacted wastes

For most treated mixed wastes in the EM inventory, one or more of the above waste forms can meet the requirements of chemical durability, for example, leach resistance and long-term stability, physical strength and fracture resistance, and resistance to radiation damage[5] (Mayberry and Huebner, 1993; Ewing, et al., 1995).

Compatibility with the waste stream is a primary consideration in selecting among the available waste forms. In some cases, lack of compatibility limits the selection. For example, organic liquids are usually incompatible with grout and can degrade the physical properties of the grout matrix even when present in small amounts. Similarly, soluble inorganic salts leach from grout and, in some cases, from polymer-encased waste forms. If compacted waste containing paper, clothing, rubber gloves, and other materials is placed in the disposal facility,

[5] Performance requirements for waste forms are discussed in Chapters 5 and 6. Radiation damage can cause loss of chemical durability and physical strength. In some waste forms radiation can produce gases. Recent reviews of radiation effects from HLW in ceramics and in glass can be found in Weber, et al., 1997, and Weber, et al., 1998.

organic decay products will form gradually. Since many of these organic compounds can chelate heavy metals and radionuclides, increased mobilization of the waste compounds could result. In the following parts of this section, each waste form will be described, along with its advantages, disadvantages, and state of development.

Cement-Based Grout

Grout stabilization is now the process of choice for most of the routine mixed waste stabilization operations at DOE sites. [6] Historically, grout has been one of the most commonly used materials for solidifying and stabilizing low-level radioactive wastes, and its technology is at a mature stage of development. Grout stabilization of RCRA heavy metals is standard technology for producing waste forms that meet EPA requirements (Kalb, et al., 1997; Conner, 1990). Chemical and physical properties of concrete are well known, and experience with concrete in construction is extensive. However, to achieve optimum physical properties and leach resistance for wastes, grout forms must be formulated taking waste composition into account.

Grout commonly is used in making waste forms because of its low cost and ease in preparation. Grouting is accomplished by mixing the waste, water, cement, and additives in appropriate proportions at room temperature and letting the mixture harden. Chemical processes in the mixture include hydration to form colloids that coagulate into gels and colloid precipitates (the setting process), followed by gel drying and crystallization (the hardening phase). In addition to portland cement, a variety of such materials as limes, blast-furnace slags, and pozzolans (volcanic rock, clays, and diatomites), and fly ash will form cementitious

[6] "Cement" is a mixture of silicates and aluminates of calcium obtained by roasting a mixture of clay and limestone, which forms a solid when mixed with water. "Concrete" is a solid product that results from mixing cement, water, aggregate (small stones or similar materials) and additives that may affect such properties as the rate of solidification (setting) or strength. "Grout" is essentially concrete without aggregate. Waste generally behaves as an additive in the grout mixture and becomes chemically or physically incorporated into the solidified matrix.

solids. These materials have different chemical compositions of calcium, aluminum, and silicon oxides and provide different waste encapsulation properties. Under optimum conditions, inorganic materials are micro-encapsulated into the gels and become part of the crystal structure of the cement. Waste is protected inside the low-permeability grout mass. Properly formulated grout waste forms are physically strong. Compressive strengths suitable for construction can be attained.

Although grouts can be made to have low permeabilities, the grout matrix is relatively more porous than other waste form matrices (notably glass and ceramics, to be discussed later). Its ability to retain some of the more important waste constituents depends on their insolubility if water comes into contact with the grout. For example, the RCRA metals and TRU elements (e.g., Pu and Am) are retained very well by grout because they have low solubility in the alkaline pH range of 10-12. This is the typical pH range of groundwater that is in contact with grout because of calcium and aluminum oxides in the grout itself.[7]

Cesium and strontium, on the other hand, are soluble in alkaline groundwaters and are considerably more leachable from grout than are heavy metals. Anions such as chloride or sulfate are readily leachable, and they also affect the setting and strength of concrete. Organic wastes usually must be excluded from grout, because they interfere with the hydration chemistry, retard setting, lower the strength of the waste form, and are leachable. However, some wastes that contain organic contamination have been successfully treated by various grouting technologies (Means, et al., 1995).

For grout waste forms, the increase in waste volume caused by the cement itself and any required additives may be a factor in their selection. The cost penalty for disposing of a greater volume of waste forms that contain low waste loading can be evaluated against the cost of additional treatment to remove troublesome constituents or selecting a different matrix that is more compatible with the waste stream.

---

[7] Because TCLP is performed under acidic conditions, grout waste forms may fail this test for some RCRA metals, especially mercury. The TCLP is discussed in Chapters 3 and 5. Requirements under RCRA are described in Chapter 3.

Glass

After over 40 years of research and evaluation, glass has become the material of choice for immobilizing radionuclides contained in high-level radioactive waste. Vitrified waste forms generally are considered to be more stable and leach resistant than cement-based forms, but their production requires high-temperature processing in specialized equipment with careful control of chemical reduction and oxidation (redox) in the melter. Glass forms are desirable because they offer low release rates and mechanical and thermal stability in the near-surface disposal environment. Because most waste constituents dissolve in the molten glass at typical processing temperatures ranging from 1100 °C to 1500 °C, high waste loading is possible, and the waste is retained in the resulting matrix, even if the waste form is disturbed or mechanically damaged. The high processing temperature destroys organic materials and volatilizes some inorganic salts (e.g., nitrate salts), resulting in a volume reduction of the waste stream. Other salts, such as sulfates and some RCRA metals, do not dissolve well in the glass, and they can pose difficulties in producing a homogeneous glass product with high waste loading. The decision to use glass as a waste matrix is generally based on economics, compatibility of glass with the waste stream, and the required product performance (NRC, 1996a).

Borosilicate glass composed of 35–55% SiO_2, 10–20% alkali metal oxides (primarily sodium), and 7–20% B_2O_3 is the most well-known glass for waste stabilization (NRC, 1996a). Phosphate and aluminosilicate glasses also have been formulated for this purpose. In the process, waste and glass formers are mixed at high temperatures to produce a melt that becomes an amorphous solid when cooled. Leach resistance, compressive strength, and solubility of some waste constituents in the solid product can be modified by changing processing conditions or the composition of the glass.

Heating of the mixtures of waste and glass can be done by:

- Joule melters that heat by passing an electrical current through the glass;
- combustion melters using fuel to generate heat;

- graphite arc furnaces that generate heat by passing a current from a graphite electrode to the melt or another electrode; and
- high–frequency melters that use microwave energy or induction heating.

Even though glass is a robust matrix, the vitrification process and the quality of the resulting waste form depend on matching the waste composition with the appropriate glass-forming additives and on the operating temperatures. Key parameters such as waste loading, viscosity, melt temperature, and durability are interrelated. Higher temperature increases the solubility of practically all waste constituents in the molten glass and, within limits, increases the achievable waste loading and homogeneity of the resulting waste form, but higher temperature also increases corrosion of the processing equipment. In addition, higher temperature increases the volatilization of constituents like mercury and radioactive iodine and cesium, which must be trapped by an efficient off-gas system. This increases the expense, complexity, and amount of secondary waste. Organic materials will be destroyed at high temperatures in an oxidizing environment, but redox conditions must be controlled or else some of the less soluble constituents may precipitate from the melt. Crystals of insoluble inorganic salts or metals that may precipitate during the melting stage can affect the final properties of the glass and cause process problems, such as non-uniform heating and melter pluggage. For these reasons waste must be well characterized and the entire process thoroughly tested to assure that the vitrification system will perform as designed.

Polymer-Encapsulated Forms

Polymer encapsulation involves embedding waste materials in an organic polymer. The waste is either dispersed as a powdered solid (microencapsulation) or the waste is surrounded by a coating of polymer (macroencapsulation). Polymeric matrices, such as bitumen, polyethylene, epoxy resins, or polyesters are used in macro- and microencapsulation (Kalb, et al., 1997). The cost of preparing this waste form is generally between that of grout and glass, and it provides a high degree of retention until the polymer itself is degraded. The variety of polymer systems allows matching waste form performance to a specific waste within economic constraints.

Macroencapsulation envelops debris in a polymer block or in a thick protective polymer coating and is an accepted treatment technology for debris waste. Macroencapsulation complies with EPA's requirements for disposal of EPA-defined debris wastes and other wastes where macroencapsulation has been designated as BDAT. Macroencapsulation of radioactive lead solids in polyethylene has been carried out at Envirocare as part of MWFA's technology development program and is now operational.

Microencapsulation[8] uses polymer matrices to coat small particles of waste that have been prepared by size reduction. For example, polyethylene at 130-150 °C can be mixed with dry waste particles ranging in size from 75 microns to 3 mm and extruded into pellets. Other plastics (e.g., thermosets such as vinyl acetate and styrene) provide similar capabilities but at a considerably higher cost than polyethylene. Bitumen is used outside of the United States. Sulfur cement is a recently developed polymer consisting of 95% sulfur and organic monomer that may be especially useful for mercury.

Microencapsulated wastes are less likely to retain their dimensional stability than grout waste forms, and they generally require secondary containers to provide physical strength. The secondary containers also can act as additional barriers against release of the waste constituents. Both biological effects and radiation are known to degrade organic materials. Higher levels of radiation can produce radiolytic gases. Polyethylene appears to have reasonable stability to radiation typically associated with MLLW (Kalb, et al., 1997). Because waste constituents are not chemically bound to the matrix, grinding the sample as required by the TCLP partially destroys the effectiveness of encapsulation. This may cause microencapsulated waste forms to fail the TCLP for some wastes. As a consequence, this technology presently has limited applicability.

---

[8] The term "microencapsulation" as used above is not the same as defined in EPA regulations. EPA applies the term "microencapsulation" to debris waste that is stabilized using grout technology.

Crystalline Ceramics

Ceramic materials have traditionally been produced by firing clays or hot pressing similar inorganic materials at high temperatures. Ceramics have been considered as waste forms for high-level radioactive waste (Lutze and Ewing, 1988), however, there has been little deployment of this technology for waste management at DOE sites. Phosphate-bonded ceramics that can be made at room temperature and show promise for immobilizing mixed wastes have recently been developed with financial support of the MWFA (Wagh, et al., 1997; Singh, et al., 1998).

Synthetic rock (synroc) is an example of a traditional hot-pressed ceramic waste form consisting of three titanate ceramic phases (zirconolite, hollandite, and perovskite) into which 10-25% waste is incorporated at about 1200 °C. The waste components are trapped in the molecular structure of the crystals (better-known examples are the natural and synthetic zeolites that trap metal ions and are used as catalysts in the chemical industry). Because the ions or molecules that comprise the waste constituents must fit into the molecular cages in the minerals, the composition of the ceramic and the waste must be carefully matched. Contact with ground-water having a high chemical activity of silica can lead to phase alteration. Otherwise, this material is highly resistant to leaching.

Phosphate-bonded ceramics are formed by treating calcined magnesium oxide with monopotassium phosphate or phosphoric acid. The process does not require an elevated temperature, and it appears to be sufficiently inexpensive for practical application to mixed wastes. However, these ceramics are a new development and had not been thoroughly evaluated by the end of the committee's review period.

Ceramics are comprised of crystalline phases, some of which are similar to minerals, so their long-term leaching behavior and stability can be reasonably estimated from geologic and geochemical analogue data. Retention of waste constituents in properly formulated ceramics is very good. Radiation or biological activity is unlikely to affect the inorganic host matrix. For high-temperature ceramics, operational problems are similar to other high-temperature processes, such as vitrification or incineration. Although promising, the phosphate-bonded ceramics have yet to be demonstrated superior to grout in an integrated industrial-scale process for immobilizing mixed waste.

Vitreous Ceramics

Vitreous ceramics are a combination of glasses and crystalline solids. As waste forms, they are generally produced at high temperatures in plasma-heated systems. Such plasma systems have a greater tolerance for waste composition variability than glass melters. However, the higher process temperatures require close attention to volatilized waste constituents and off-gas system design. The mixture of crystalline and amorphous components in a vitreous ceramic makes assurance of product quality and durability more difficult than for homogeneous waste forms.

As discussed previously, the apparent advantage of plasma processes to produce vitreous ceramics from a wide variety of wastes led to the MWFA's interest in their potential use for heterogeneous and debris wastes and for wastes that do not form glass readily. While demonstrations of this technology were considered successful by the MWFA, the process required optimization for each waste stream. The need for waste characterization could not be avoided to the extent hoped, and therefore the process showed no overall advantage compared to the more established technologies for making grout or glass waste forms (John Kolts, personal communication to NRC staff, October 1998). Although plasma-heated systems may have applications at some DOE sites, their widespread deployment for mixed waste treatment is no longer being pursued by the MWFA.

Compacted Debris

For slightly contaminated solid waste, compaction with a hydraulic press can provide large volume reductions and produce waste forms suitable for disposal. In simple systems, wastes are compacted by pressing them into a 55-gallon drum. More powerful compaction equipment, which can easily crush metallic scrap and entire 55-gallon waste drums, is commercially available. In the latter case, several crushed 55-

gallon drums can be placed in an overpack drum of about 80-gallon capacity. Large quantities of the debris group are planned to be compacted with supercompactors to achieve large volume reductions. The compacted material will be placed in overpack drums for disposal. Some metallic debris, such as radioactive metal shielding, drums, and process equipment, will be decontaminated by cleaning the surface, compacting, and macroencapsulating. This process has been applied at Envirocare to contaminated lead ingots and will be used at the INEEL Advanced Mixed Waste Treatment Project for all debris waste.

TECHNOLOGY NEEDS

As a part of the process of developing its technical baseline report, the MWFA has identified areas where technologies necessary for mixed waste treatment are deficient or missing. These needs are derived mainly from input from the Site Technology Coordinating Groups (STCG)[9] at each DOE site. The input was prioritized by MWFA and presented in the Technical Baseline Report (DOE, 1996a, 1997a). The MWFA list of 24 technology needs is reproduced in Appendix C. The top four needs in the prioritized list are important to MWFA's task of assuring that adequate waste forms are available for DOE's mixed waste inventory.

The first priority is waste characterization. It was pointed out in Chapter 2 that the committee considers quantitative knowledge of the EM mixed waste inventory to be deficient, and that definition of detailed flowsheets for waste treatment and stabilization is not possible without this information. If waste to be treated is not well defined, only the most robust processes, such as vitrification or formation of vitreous ceramics, can be selected with confidence, and the process must be designed for conservative (possibly low and less efficient) waste loadings, thus increasing the volume of waste and cost of disposal.

Mercury and salt stabilization are the second and third priorities. Most treatments described in the previous section include provisions for

---

[9] A group was formed at each DOE site to assist EM in identifying site technology development needs. STCG members include representatives of DOE, contractors, EPA, tribal nations, and other stakeholders.

removing mercury from the primary or secondary waste stream. However, these methods to remove and stabilize mercury require further testing and development. In addition to mercury stabilization, there are four other mercury-related technology needs on the list. The difficulty of stabilizing inorganic salts was also described in the previous section. Salts cause problems with most waste form matrices, including grout, glass, and polymers. Presently, these problems can be overcome only by reducing the waste loading in these waste forms. A similar need to improve waste loading for incinerator ash is listed as priority 10.

The fourth priority is the need for assessing the behavior of waste forms in the disposal environment. Without an objective, defensible means to evaluate waste form performance in the disposal environment, waste managers may be required to use the most advanced and expensive waste forms available. Conversely, in the absence of tests to assess the waste forms, disposal facility designs may take no credit for the waste form and rely entirely on other features of the disposal site, including engineered barriers and site attributes to assure safety. The topics of waste form characterization and performance assessment are discussed in Chapters 5 and 6 of this report.

FINDINGS, DISCUSSION, AND RECOMMENDATIONS

In following its statement of task, the committee gave special attention to assessing the state of technologies that MWFA has selected or developed for treating EM's inventory of mixed waste and producing waste forms for disposal. Based on MWFA's presentations to the committee (Appendix B), information in the section on Available Waste Forms, and the judgments of its members, the committee found that available classes of treatment methods and waste forms are sufficiently developed to accommodate DOE's current and expected mixed waste inventory. There is, however, a continuing need for improved engineering adaptation of these technologies to the actual mixed waste streams they are intended to treat.

The committee found the following:

- EPA hazardous waste regulations have been a major driver in technology selection and development by the MWFA. Other drivers (e.g., economics) have received less attention in MWFA's programs.
- Physical properties of the waste (e.g., solid, liquid, debris) or broad chemical properties (e.g., combustibility) have been the basis for selecting treatment technologies through the categorization of EM's mixed waste into five treatment groups. This has provided an efficient means of defining generalized flowsheets for waste treatment and identification of generally compatible waste forms, but it is not sufficient for engineering design and optimization of treatment processes, nor does it reflect the hazards posed by the various portions of the inventory.
- The selection of waste forms has been linked primarily to their compatibility with proposed treatment processes, rather than factors such as the disposal environment. The committee recognizes that this is because of, at least in part, a lack of realistic tests of long-term waste form performance in the disposal environment, and the minor role assigned to the waste form in current performance assessments. These factors will be discussed in Chapters 5 and 6.
- MWFA has recognized the need for better characterization of EM's mixed waste inventory, but it has not explicitly addressed the trade-off between detailed characterization and robust treatment processes that could accept more heterogeneous waste streams.
- MWFA has not given sufficient attention to the engineering work necessary to adapt existing or new technologies to operation with radioactive materials and to demonstrate these technologies on a production scale.
- Privatization is emphasized in EM's planning for mixed waste management. Even with privatization, technology development will still be required where there are deficiencies in available treatment or waste form technologies. The division of responsibility for technology development among MWFA and

contractors is not clear, nor are the mechanisms for sharing results of technology development efforts well defined.

MWFA presentations and reports reviewed by the committee show that there are ample treatment technologies and waste forms to accommodate the wide variety of mixed wastes in the EM inventory. Although some wastes categorized as unique may require development of specific treatment methods, the committee believes that existing classes of waste forms are suitable for these wastes.

The limited inventory data that are available, as discussed in Chapter 2, suggest that treatment of solvent- and mercury-contaminated wastes should receive priority. Technologies for treating these wastes exist but need considerable development in application to mixed wastes. Optimization of existing waste form technologies to allow higher waste loadings and provide less expensive production methods certainly is possible, and potential cost savings justify continued effort toward process optimization. In particular, higher waste loadings (e.g., ash and salt) would reduce the volume of waste to be disposed and the concomitant disposal costs.[10] Mercury stabilization is one of the MWFA's top priorities for technology development.

Better characterization of EM's mixed wastes can reduce uncertainties in the composition of the waste streams to be treated. This in turn should allow design of simpler treatment processes that accept a more narrow range of waste compositions. The process should be less expensive to build and operate than those that must accept poorly characterized wastes. Processes that treat well-characterized waste streams can be optimized to produce higher quality waste forms and less secondary waste.

The committee noted that, whereas many valuable treatment technologies have been identified by the MWFA, development steps were sometimes bypassed and technology deployment has not always been successful. Two examples, in the committee's opinion, were molten metal technology and the plasma torch. The committee viewed the plasma torch as an advancement over vitrification due to its potential to treat wastes of widely varying compositions. In spite of the major

[10] Disposal costs are usually based on waste volume.

technical difficulties encountered in implementing this technology at INEEL, the committee believes that a more methodological, stepwise development program and more careful adaptation to real wastes might have led to successful deployment of the plasma torch. If properly developed, the plasma torch could play a useful role by treating a portion of the EM mixed waste inventory that is poorly characterized.

Molten metal technology appeared to be a case where private vendors promoted development of a technology expected to be similar to the plasma torch in its applicability. Molten metal technology, while applied in a limited scope (EPRI, 1997a; Evans, 1997), was not pursued by MWFA. A planned visit by the committee to the molten metal demonstration at Oak Ridge was canceled by DOE, and the committee received no first-hand information about the practical results of the demonstration tests.

The promise of improved process technology must be weighed against the time frames (milestones) mandated in regulatory agreements, the overall goals of EM's "Paths to Closure" (DOE, 1998c), and cost of technology development. The many steps between identifying a new technology and its deployment include demonstrating safe operation, economic viability, and effective processing capability with actual waste streams, as well as acceptance by site operators. The committee cautions that problems in technology transfer may arise if a technology developer simply hands off new technologies to the user. It is the experience and judgment of the committee that considerable interaction with the user and support by the technology developer will be necessary to assure successful implementation.

EM expects privatization to yield cost savings, schedule acceleration, and other advantages for many of its cleanup projects (DOE, 1998c). Although assessing privatization was not within the committee's task, plans for the Advanced Mixed Waste Treatment project were presented to the committee and the committee learned that EM expects a large fraction of the mixed waste inventory to be treated by private contractors. Privatization may offer advantages in managing and implementing complex projects. However, the extensive reliance on private contractors to design and operate mixed waste treatment facilities raised some concerns among committee members.[11] One concern of the committee is how technologies that have been established by years of operating experience in DOE can be transferred successfully to a private contractor having relatively limited experience with DOE wastes. Another concern is how to assure that adequate knowledge of the characteristics of mixed waste to be treated is available to a bidder. A third concern how a contractor may trade-off "best" technology versus "minimum acceptable" technology.

Recommendations

The committee's general recommendation is that MWFA should no longer emphasize the development of new classes of waste forms. After reviewing the technologies available to treat EM's mixed waste inventory, and considering the resulting waste forms, it is the committee's judgment that no new classes of waste forms are required. Clearly no single form is appropriate for all wastes, but, collectively, the variety of available waste forms and well-established waste form production technologies make it unlikely that any totally new class of waste forms will be necessary to complete EM's planned cleanup program. Grout waste forms, for example, can accommodate essentially all mixed wastes in the inventory, although pre-treatment is required for some wastes (e.g., organics) before grouting. Where grout may be inadequate for either technical or regulatory reasons (e.g., failure of the TCLP), glass or polymers can be used. As discussed in this chapter, the advantages and disadvantages of each of these principal waste forms and the technologies for making them are well known. Vitreous ceramics comprise another well known class of waste forms that could accommodate most or all of the inventory, but at the present time the MWFA has no established technology for making these forms.

MWFA should now emphasize the engineering design, integration, and scale-up of its proposed treatment processes and their demonstration and deployment, as needed, at the DOE sites. Technology development and deployment must consider the overall EM waste

[11] The committee was aware of the failure of the Pit 9 privatization contract at INEEL.

management strategy that is described in "Paths to Closure" (DOE, 1998c).

The committee recommends the following:

- MWFA should integrate individual treatment technologies being developed for its five treatment groups into an overall mixed waste management system.
- MWFA should demonstrate new treatment technologies on at least the pilot plant scale using real wastes or realistic surrogates before the technology is designated as ready for deployment.
- MWFA should continue to address technology deficiencies that it has identified through input from the Site Technology Coordinating Groups and update its Technical Baseline Report to reflect progress in addressing these deficiencies.
- MWFA should continue to provide research funding for developing robust processes such as the plasma torch that can treat and stabilize waste of poorly defined or variable composition.
- MWFA should continue basic research related to the understanding of the physical and chemical attributes of waste forms.

5

Characterization of Mixed Waste Forms

The possibility for effects on human health and the environment from mixed waste disposal depends on the rate at which the waste constituents are released from the waste form and the concentration of these materials as they move into the environment. Quantitative information about the mechanisms and kinetics of waste form degradation and contaminant release during very long time periods under conditions expected for the disposal facility can provide a scientific basis for estimating these possible effects. Characterizing the chemical and physical properties of waste forms through laboratory and field testing is an essential first step in understanding their ability to control the release of contaminants.

During the past 15 years extensive research has been conducted in the U.S. and worldwide to develop methods to determine the physical and chemical integrity of waste forms, contaminant release mechanisms, and waste form degradation rates (Cunnane, 1994). Much of this research has been sponsored by the U.S. Department of Energy (DOE) in support of its high- and low-level radioactive waste management programs and by the U.S. Environmental Protection Agency (EPA) to ensure that waste forms comply with Resource Conservation and Recovery Act (RCRA) requirements. This chapter describes methods that have been used to characterize the physical and chemical behavior of stabilized waste forms. This information is needed to determine a waste form's suitability for disposal and its long-term behavior in the disposal environment.

Although there are no long-term stability criteria for stabilized mixed wastes, U.S. Nuclear Regulatory Commission (USNRC) regulations require the evaluation of disposal facility performance for periods up to 10,000 years, as described in Chapter 3. From a waste management perspective, the fundamental question is not whether a waste form will decompose during such a long time period, but when and at what rate. Short-term failure could result in a high concentration (pulse) of contaminants reaching the environment, whereas slow failure occurring over hundreds or thousands of years would produce insignificant effects.

Determination of performance of the waste form by tests lasting generally a few months and then extrapolated to estimate performance over centuries or millennia introduces considerable uncertainty into predicting long-term performance. The possibility of early failure of the waste form with subsequent generation of contaminated leachates in the disposal facility places an extra burden on design of the engineered and natural barriers and emphasizes the selection of disposal sites offering favorable geology, hydrology, and climatic conditions.

In addition to estimating long-term performance, there are a number of reasons for determining the chemical and physical characteristics of waste forms. These include (Franz, et al., 1994):

- Regulatory compliance: Show that the waste meets criteria established by state and federal regulatory agencies prior to disposal.
- Waste form development: Testing to measure and compare the effects of the many possible modifications of waste form composition and fabrication to identify the best formulation to satisfy specified criteria.
- Waste form comparison: Testing to compare the performance of alternative waste forms to assist in selecting a suitable matrix for a particular application.
- Data for site performance assessments: Tests to produce data for use in predicting the long-term performance of a disposal facility. and
- Quality control: Rapid, routine testing during waste processing.

The procedures and data needed to satisfy each of these needs can be quite different. For example, testing procedures for a quality control program must be relatively inexpensive, suitable for repeated application, and provide rapid results. In contrast, testing procedures used to develop data for a waste form qualification or for a site performance assessment (PA) involve estimating long-term stability of the waste form as described earlier. Such procedures may involve lengthy tests and be very expensive.

Four factors are important when selecting tests to characterize a waste form (Poon, 1989):

- representation (of actual field conditions);
- compatibility (same test applicable to different types of wastes);
- reproducibility (consistent results from repeated testing); and
- test stability (consistent results as a function of the duration of testing).[1]

The most common approach to estimating long-term behavior of a waste form is to evaluate its performance under severe conditions over a short period. For cement-based waste forms, the most common means of accelerating testing is through use of higher water flow rates and more aggressive leaching solutions (Quillin, et al., 1994). As will be seen in the next section, this approach has led to the design and implementation of leaching tests that bear little resemblance to the environmental conditions experienced by the disposed waste. There is an increasing recognition of the need for tests that elucidate reaction mechanisms and that determine reaction rates over a range of repository-relevant conditions, and, if possible, to confirm results of laboratory investigations with data from analogues, for example ancient natural or man-made glasses (NRC, 1996a).

Franz, et al. (1994) proposed that tests used to evaluate the performance of solidified mixed low-level waste (MLLW) be organized

---

[1] The committee recognizes that there are difficulties in practice with applying each of these factors. For example actual field conditions may not be predictable for the long-term, and poorly designed experiments can appear to give consistent results.

into two tiers, which are summarized in Table 6. In the first, the leachability of a waste form is measured, because low leachability is the most important single disposal criterion. If a waste form cannot sufficiently limit the leachability of hazardous or radioactive constituents, it will almost certainly not be considered suitable for disposal in an MLLW facility.[2] Only wastes that meet the specifications of appropriate leach tests should be subject to the second tier of testing. The second tier of testing is designed to determine if conditions in a disposal facility will affect the integrity of the waste form and, therefore, its ability to retain hazardous and radioactive contaminants. These two testing tiers, tests for leachability and the physical durability of waste forms, are discussed in this chapter.

Along with waste form characterization for directly quantifiable parameters (e.g., leach rate, corrosion rate, and mechanical properties) much effort has been spent in the past on more fundamental properties such as physical structure, diffusion phenomena, impact of radiation and biochemical phenomena. The latter information enhances understanding and interpretation of directly measured characteristics.

LEACHABILITY

The potential for leaching waste components from a waste form depends on external factors and the intrinsic characteristics of the waste form. These include the following:

- chemistry of the leaching fluid (lixiviant) and the near-field barriers;
- hydrodynamic regime near the waste (i.e., the rate of lixiviant flow past the waste);
- chemical properties of the waste form matrix and the waste constituents;
- physical properties of the waste form, particularly its porosity and specific surface area; and

---

[2] In the special case of mixed transuranic (MTRU) waste destined for disposal in the Waste Isolation Pilot Plant, leachability is not a consideration.

TABLE 6 Summary of Testing Procedures Suggested for MLLW Waste Forms

First Tier—Applicable to all candidate waste forms

- Leachability
  - Release of radionuclides
  - Release of RCRA constituents

Second Tier—Applicable to waste forms that pass leachability tests

- Compressive strength/non-destructive testing (NDT)
- Stability in water
  - Dimensional changes
  - Compressive strength/NDT
- Stability after irradiation
  - Dimensional changes
  - Leachability
- Stability after freeze/thaw cycling
  - Dimensional changes
  - Leachability
- Stability after wet/dry cycling
  - Dimensional changes
  - Compressive strength/NDT

SOURCE: Franz, et al., 1994.

- biochemical properties of the waste form and the disposal environment.

The leaching rate is important since it will determine the concentrations of hazardous or radioactive constituents that can potentially move into the environment. A summary of some of the leaching tests that are most relevant to mixed waste forms is presented in Table 7.

TABLE 7 Important Leach Tests for Hazardous and Radioactive Waste Forms

Test	Leaching Solution	Duration	Comments
Toxicity Characteristic Leaching Procedure (TCLP)	Acetic acid	18 hours	Leaching of crushed sample used to define characteristic hazardous wastes
ANS-16.1	Deionized water	Up to 90 days	Measures leach rate from monolithic waste form
Accelerated Leach Test (ALT)	Deionized water`	11 days	Elevated temperature and leach solution changes used to measure leach rates

SOURCE: 40CFR 261, Appendix II, Method 1311; ANS, 1986; Fuhrmann, et al., 1990.

The Toxicity Characteristic Leaching Procedure (TCLP) is required by EPA for hazardous wastes regulated by RCRA.[3] The procedure was developed to simulate typical conditions in a municipal solid waste landfill with a high concentration of organic acids resulting from decomposition of organic waste materials. The TCLP uses a buffered acetic acid solution mixed with crushed waste at a liquid-to-solid ratio of 20:1. After the sample is agitated in a rotary tumbler for 18 hours, the leachate is filtered and analyzed for regulated constituents. By requiring the waste form to be crushed and then leached with an acidic lixiviant, the TCLP subjects the waste form to more severe conditions than are likely in a properly designed disposal facility. Although recognizing that data from this single-point test are not sufficient for estimating the waste form's long-term performance in an actual disposal facility, the EPA considers the procedure to be a practical way that the many different waste types and forms subject to RCRA can be qualified for disposal.

As a means for demonstrating acceptable leach resistance of

[3] RCRA and the requirement for use of the TCLP are discussed in Chapter 3.

solidified Class B and C radioactive wastes, the USNRC recognizes a test developed by the American Nuclear Society (ANS, 1986), the ANS-16.1 procedure. Rather than using crushed waste as is done in the TCLP, the ANS-16.1 procedure specifies that a monolithic cylinder be leached with deionized water at a volume-to-surface-area ratio of 10 cm. The water is sampled and replaced at 2 hours and 7 hours, and then at 1, 2, 3, 5, 14, 28, 43, and 90 days. This test therefore measures leaching as a function of time, in contrast to the single measurement that is made with the TCLP. The ANS test includes calculation of a "leachability index" that can be related to an effective diffusion coefficient for waste forms that leach mainly by diffusion (rather than by dissolution of the waste form matrix itself) during the test period.

The ANS-16.1 procedure also is intended to provide quality assurance information during the production of waste forms and to make rapid intercomparisons of waste forms in laboratory testing. This information is provided by chemical analysis of the first few samples (e.g., after 2 and 7 hours), which will reasonably quickly determine whether the quality of a tested waste form is inferior to that of others tested previously. For quality assurance and product intercomparison purposes the test does not require any assumptions about the leaching mechanism.

Investigators at Brookhaven National Laboratory (Fuhrmann, et al., 1990) have developed an Accelerated Leach Test (ALT) intended for higher leach-rate materials, such as portland cement, in which leaching is controlled by diffusion in the porous medium. It is similar to the ANS-16.1 procedure because the leaching solution is sampled and replaced periodically. Elevated temperatures, large volumes of lixiviant, frequent lixiviant change, and small specimen size are used to accelerate contaminant release. The test can be completed in 11 days, and the results extrapolated to 20 °C to allow determination of an effective diffusion coefficient. Analysis of the data also can indicate whether the contaminant release is controlled by diffusion or some other process, such as dissolution of the waste form itself. Both the ANS-16.1 and the ALT methods yield effective diffusion coefficients, which make their methods more appropriate for estimating long-term behavior of the waste form and for input into a PA.

The American Society for Testing and Materials (ASTM) has developed a number of tests that can be used to characterize waste forms:

- ASTM Method D 4874-95, "Standard Test Method for Leaching Solid Material (Waste) in a Column Apparatus;"
- ASTM Method D 5233-92, "Standard Test Method for Single Batch Extraction Method for Wastes;"
- ASTM Method D 5284-93, "Standard Test Method for Sequential Batch Extraction of Waste with Acidic Extraction Fluid," for wastes with at least 5% solids; and
- ASTM Method D 5369-D, "Standard Practice for the Extraction of Solid Waste Samples for Chemical Analysis Using Soxhlet Extraction" at elevated temperature with cycling organic solvent for the extraction of non-volatile and semi-volatile organic compounds.

A more recent ASTM method, C1308-95, is an "Accelerated Leach Test for Diffusive Releases from Solidified Waste and a Computer Program to Model Diffusive, Fractional Leaching from Cylindrical Waste Forms." This method accelerates the leach rate and determines if leaching is diffusion controlled. The resultant data may be appropriate for use in a PA.

Many other characterization tests were developed in other laboratories in the U.S. and abroad. They all tend to meet one of the two major objectives:

1. provide information for performance analysis
2. prove conformity with quality standards, regulations and waste acceptance criteria.

Only a few of the presently available leach tests provide a basis for evaluating long-term behavior. Solution compositions are determined, but the identification of the phases that form as a result of the alteration generally is not required in the test. The latter information is essential to determining reaction progress and evaluating long-term behavior. With regard to characterization for performance assessment, increased attention is being given to testing in realistic conditions (e.g., in site-specific groundwater).

DURABILITY

This section summarizes methods of estimating the long-term durability of proposed waste forms. There is an extensive body of literature on this topic, much of it generated in support of high-level nuclear waste disposal programs in the United States, Europe, and Japan (Lutze and Ewing, 1988). The following discussion is limited to the two waste forms most frequently considered for MLLW and MTRU: cement-based grouts and glass. Four activities for predicting the long-term durability of waste forms are laboratory testing, field testing, analogue studies,[4] and modeling (Means, et al., 1995).

Laboratory testing involves developing procedures for accelerating the effects of degradation processes. These processes include thermal cycling, radiation damage, biological degradation for wastes containing biodegradable compounds, dissolution, and structural failure (Mayberry, et al., 1993). Means, et al. (1995) provide a brief review of laboratory testing.

Although a successful waste form must maintain its physical integrity over long time periods, the short-term laboratory procedures available to assess durability have not been demonstrated to replicate field behavior (Kirk, 1996). According to Kirk, research must be undertaken to:

- provide equivalent test parameters to the physical and chemical forces in the field; and
- calibrate the laboratory time scale to long-term field exposures, including cyclic conditions, elevated temperature, and

---

[4] In its survey of methods to characterize waste forms, the committee did not review studies of analogues. These studies of ancient concretes and glasses as well as similar natural analogues have been conducted to better understand the long-term behavior of these materials in the environment. Several studies of natural glasses are presented in NRC, 1996a.

chemical conditioning that would be equivalent to long-term repository conditions.

Franz, et al. (1994) identified needs in the existing tests that currently are employed to characterize MLLW forms and the absence of tests they feel are required for the evaluation of MLLW forms, especially when predicting long-term behavior. For grouted wastes these needs include:

- an improved thermal cycling test;
- a means to judge pass or fail for wet and dry testing;
- a procedure for irradiation of waste forms, including definition of test conditions (e.g., dose rate, temperature, and moisture);
- a non-destructive test to replace compressive strength measurements so that the same sample can be tested repeatedly; and
- pass or fail criteria for dimensional changes and loss of material by spalling.

An example of the application of models is the work by Atkins, et al. (1994), who investigated the performance of cementitious waste forms by considering the chemistry of the calcium-aluminum-silica phases present in ordinary portland cement. Using this approach, they were able to investigate the effects of elevated temperature, pH, and cement and ground water interactions. In particular, they were able to evaluate the effects of waters containing high concentrations of sulfate, chloride, and carbonate on a cementitious waste form. Lee, et al. (1995) adopted an integrated approach to modeling the long-term durability of concrete in a disposal facility that included a model of the concrete and calculation of pore fluid speciation, coupled with mass transport in and near the concrete. The models were used to simulate degradation of concrete over a period of 300 years. They found that the alkali elements (sodium and potassium) control the pore chemistry of the concrete for much longer than most previous barrier degradation studies assumed.

There are few field studies of actual waste form performance principally because of the lack of facilities in which stabilized waste could be placed. Means, et al. (1995) cite only one study in which grouted waste was sampled after 9 and 18 months of burial. Testing showed that lead and other metals remained immobilized, while the physical properties and the porosity of the waste form decreased slightly. The organic contaminants, however, were not immobilized effectively.

Extensive testing protocols for vitrified waste forms have been developed to support high-level waste disposal initiatives. Long-term glass stability depends in large part on maintenance of a silica-saturated solution around the waste form. The effects of variations in lixiviant composition can readily be measured in lab tests. Usually these tests must be run over a long period of time, making them expensive to conduct. Laboratory simulation of long-term glass stability frequently includes studies at elevated temperature. Higher temperatures affect the solubility of glasses and the kinetics of release mechanisms. These higher temperature effects must then be correlated to leaching and degradation rates at temperatures likely to be encountered in a repository. A second factor that has been investigated extensively in glasses is the effect of radiation damage (NRC, 1996a; Weber, 1997).

FINDINGS, DISCUSSION, AND RECOMMENDATIONS

To determine if current waste forms are sufficiently developed to stabilize EM's inventory of mixed waste, test methods to characterize the waste forms must be available. The committee reviewed the methods available to characterize the chemical and physical stability of waste forms for mixed waste. The committee found that no test methods are accepted by the technical and regulatory authorities to demonstrate the long-term (greater than a few hundred years) behavior of a waste form in the disposal environment. Available test methods can be used to measure the short-term stability of the waste. Because some mixed wastes contain long-lived radionuclides and chemically hazardous constituents, knowledge of the long-term behavior of waste forms is necessary if credit is to be taken for the waste form in assessing the long-term performance of disposal facilities.

Committee findings include the following:

- The different regulatory approaches taken by EPA and USNRC, as discussed in Chapter 3, lead to different approaches to waste form testing. This is manifested by pass-fail waste form tests, such as the TCLP that are required by EPA, and tests that provide data for performance assessment of the disposal system, as required by USNRC.

- The long-term behavior of a waste form in a disposal facility is difficult to assess because there is no agreed way to extrapolate laboratory tests conducted for periods of days to years to behavior over hundreds or thousands of years. It is difficult to elucidate the reaction mechanisms pertinent to waste form degradation that can be used to estimate long-term performance. Laboratory methods to accelerate natural processes that govern waste form degradation usually subject waste forms to unrealistic conditions.

The best that can be expected of any disposal facility is that it will release its inventory of very-long-lived radionuclides and hazardous materials slowly over long times so that their effects in the environment are inconsequential and within regulatory limits. The available waste forms discussed earlier in this report can play a very beneficial role in retarding the release of waste constituents. After they have come into contact with ground water, most good quality waste forms can be counted on to retard the release of waste constituents for long periods of time, typically hundreds of years. The relevant question regarding the behavior of a waste form in a disposal facility is not if it will fail, but when and over what period of time. The committee therefore recognizes the value of efforts to characterize waste forms in terms of fundamental physical and chemical factors that govern their stability and eventual degradation, including physical structure, diffusion phenomena and, where appropriate, effects of radiation and biological activity.

Because the committee considered the problem of characterizing the long-term behavior of waste forms to be relevant for all of DOE's wastes, not only mixed waste, the committee's recommendations are directed toward OST, which is primarily responsible for research and development in EM. The committee recommends the following:

- OST should continue to support programs aimed at fundamental understanding of waste form durability and degradation processes. These programs should lead to a better representation of the waste form in PA modeling.
- OST should work with EPA and the USNRC to develop criteria for rapid testing protocols that can be used to determine whether a stabilized waste is suitable for disposal and to assist in quality assurance and quality control in the waste treatment and stabilization process. The objective of this rapid testing protocol would be to reduce the need for performing TCLP analyses on every batch of waste prior to its disposal.
- OST should work to promote consensus among EPA, USNRC, DOE, and the scientific community on waste form testing protocols that are generally acceptable for providing at least a qualitative evaluation of long-term waste form performance in disposal environments.

6

Role of the Waste Form in Performance Assessment

The long-term behavior of a waste disposal facility is a function of the entire disposal system, including the waste form, engineered barriers, and surrounding environment. In order to assess the ability of a given disposal concept to meet regulatory requirements it is necessary to consider the influence of each of these system components on short- and long-term performance. This is accomplished through the performance assessment (PA) process. As discussed in Chapter 3, in order to permit a low-level waste (LLW) facility, both U.S Department of Energy (DOE) and U.S. Nuclear Regulatory Commission (USNRC) regulations require completion of a PA. The PA is the only step in the mixed low-level waste (MLLW) management process the where long-term stability and performance of the waste form is part of the formal evaluation of a disposal system. Ascertaining the role of the waste form in current PA methodology was necessary for the committee to evaluate the adequacy of available waste forms to meet present regulatory criteria. This chapter gives an overview of current PA methodology, the role of the waste form in PA, and the committee's findings and recommendations for improving the waste form's representation in future PA methodology.

Performance assessment is based on a mathematical model of the proposed facility and its environment. Results of this modeling can be used to help demonstrate that a disposal facility will protect the health and safety of the public. The PA process addresses all potential exposure pathways that may occur at the site and allows comparison of the doses estimated from performance calculations with the performance objectives in DOE Order 5820.2A to determine whether the proposed facility will be in compliance. As noted in Chapter 3, DOE Order 5820.2A requires that a PA be prepared for each proposed DOE disposal facility for LLW and MLLW. No waste form performance criteria have been established solely for mixed waste, therefore, these wastes are subject to the requirements for LLW established by the USNRC and to the leachability criteria established for hazardous wastes by the U.S. Environmental Protection Agency (EPA). Recently, the USNRC has provided guidance for the preparation of PAs for LLW disposal that addresses the issue of waste form (USNRC, 1997).

This chapter describes the role of the waste form in the PA methodology developed for LLW disposal facilities in the United States.[1] Emphasis is placed on the rate of release of waste constituents from stabilized waste in conjunction with other physical and chemical processes that can affect future exposure to humans and the environment. It is important to recognize that a PA is only required for the radioactive constituents of a waste and not for the chemically hazardous materials, as defined by the Resource Conservation and Recovery Act (RCRA), that are present in mixed waste.[2] Although substantial work is being done on developing probability based models for risk assessment of hazardous materials that consider variability and uncertainty, they generally do not extend over the long time periods required by the USNRC for its assessment of disposal facilities.[3]

Performance assessment encompasses the entire disposal facility as an integrated system. It involves constructing a conceptual simplification of the system (conceptual model) that can be represented with mathematical models of the process involved. The mathematical models

---

[1] Performance and risk assessment methodologies differ among countries due to differing regulatory approaches. For example, in some European countries, the time horizon for PA may extend beyond that required in the U.S., with results for longer times becoming more qualitative. Instrumental regulations for disposal of radioactive and hazardous wastes are being harmonized, with PA likely to include disposal systems for both types of waste (Seitz, 1998).

[2] See Chapter 3 for a discussion of RCRA.

[3] See, for example, Amendola (1992) and Shevenell (1993).

are then used to calculate the exposure to a critical group (a hypothetical group of maximally exposed persons at a specified location near the facility) as a result of releases from the facility. The waste form itself is considered to be one of several barriers that act to prevent or retard the release of radionuclides. Quantitative data are needed to describe the types of waste that will be received, the components of the disposal facility (especially the liner and cover designs), and the site characteristics. Because all of the needed data usually are not available, the models typically include estimated parameters that are derived from assumptions about the system that cannot be validated experimentally. In addition to long-term characteristics of the waste form, these parameters may include the following: (1) infiltration rates of water into the facility, (2) time to failure of the engineered barriers, (3) water flow through the unsaturated zone, (4) ground water flow rates, (5) interaction between the contaminants and soils, (6) atmospheric mixing, and (7) probability of inadvertent intrusion. Each of these parameters can have a significant effect on the results of the PA calculations.

The USNRC (1997) notes that the goal of the PA analysis is not to predict the future but rather to test the robustness of the disposal facility against a reasonable range of future scenarios. For time periods extending to 1,000 years and beyond, there is considerable uncertainty associated with site conditions due to potential processes such as climate change, seismic events, and volcanic activity. The effects of these uncertainties are captured in the sensitivity analysis performed as part of the PA calculations. Sensitivity analysis, the identification of the parameters that, when changed, can have a significant effect on the conclusions of the assessment is an essential part of the PA process.

Mayberry, et al. (1993) have noted that there has been little feedback between laboratory testing programs and development of PA methodologies for LLW disposal.[4] Without detailed understanding of release mechanisms, laboratory data cannot be confidently extrapolated to the long time periods and environmental conditions modeled in PA. Thus, conservatism is often introduced into waste form-related calcu-

---

[4] The committee recognizes that there have been some cooperative efforts among laboratory scientists and PA modelers at several DOE sites that have resulted in better representation of the waste form in PA, for example Sullivan, 1994.

lations with the result that they may be overly conservative. There is very limited experience with actual field data from disposal facilities containing stabilized waste forms.

The release of contaminants from a waste form is the initial step in any exposure pathway, and is a major factor in the performance of the disposal facility, although it is the overall system that is assessed for compliance with the performance objectives. The waste form is often referred to as the source term. There are three general pathways that may result in exposure to radioactive or hazardous compounds: (1) aqueous transport, (2) airborne transport, and (3) inadvertent intrusion into the facility. These are discussed in the following sections.

AQUEOUS TRANSPORT SCENARIOS

The waterborne pathway is the most widely recognized and generally the best quantified exposure pathway in considering disposal of solid waste materials, including hazardous and radioactive wastes. This is in large part because of the long experience with municipal solid waste landfills and with RCRA hazardous waste landfills. Instances of ground and surface water contamination from landfills are widely known and have been studied for many decades. Indeed, most of the site selection and design criteria for both sanitary and hazardous waste landfills are oriented towards preventing water contamination.

Waterborne exposure pathways involve a multi-step process in which hazardous or radioactive constituents leach from a waste form are transported through an engineered barrier (e.g., a liner) if it is incorporated in the facility design, out of the disposal facility to the surrounding geologic strata, and finally through the unsaturated and saturated geologic formations around the site to a critical group. The principal role of the waste form in this sequence is to limit the rate of leaching of the contaminants.

Radionuclide release from a waste form may result both from physical processes and chemical processes. These include simple wash-off and more complex processes such as dissolution, diffusion, and sorption/desorption. Wash-off is the result of aqueous contact with surface contamination on an insoluble surface, such as plastic, metal, or glass. Dissolution is primarily associated with long-term leaching of stable waste forms, whereas diffusion release occurs from porous waste forms over a shorter time. Desorption occurs when a contaminant is alternatively bound on a surface (e.g., a soil particle, the waste form, or degradation products from the waste form) or released into solution in response to a reversible, quasi-equilibrium chemical process. The equilibrium is modeled using a linear distribution coefficient, or K_d, which represents the equilibrium between the sorbed constituents and those in solution. [5] Recognizing the different release mechanisms is important in developing a conceptual model for a PA scenario.

In addition to these release mechanisms, possible limits on maximum aqueous concentrations of some constituents in the leachate may occur because of solubility constraints. Solubility will be determined by the complicated chemistry and biochemistry of the leaching solution and the waste form. For example, even though a contaminant might be rapidly released in laboratory experiments with grouted waste, the high pH associated with the grout might cause this contaminant to precipitate in the leachate in a disposal facility, thus, limiting the overall release rate. Identifying which mechanisms control the rate of contaminant release from a stabilized waste may require extensive chemical and theoretical analysis, hence, it is common to simply report an overall release rate that is specific to the waste form and leaching solution.

With respect to the waste form's impact on the PA calculations, it is important to note that the release rates measured in laboratory tests are generally conservative in that they expose the waste form to far more water than would be experienced in a properly sited and constructed disposal facility. Furthermore, there is a much higher degree of mixing in laboratory tests than in a disposal facility, which acts to further increase effective diffusion coefficients. However, it should also be noted that release rates measured under laboratory conditions do not take into account heterogeneities that are likely to occur in actual waste forms produced by full-scale treatment and stabilization processes.

The USNRC does not provide universal guidance on the length of time that a disposal facility can take credit for waste immobilization

---

[5] K_d is referred to as the sorption coefficient in descriptions of ion exchange processes. It is the concentration of a given ion on a solid sorbent divided by the concentration of that ion in the ambient solution.

by a particular waste form. However, in its guidance document (USNRC, 1997) the USNRC notes that the performance of the waste form has very little effect on the containment of long-lived radionuclides (i.e., isotopes with half-lives greater than about 30 years). Unless special features are incorporated in the facility design, the USNRC recommends assuming that engineered barriers (including waste forms and containers) will physically degrade after 500 years. In the degraded condition, a barrier can still function to limit migration of contaminants, however, its performance will be based more on the chemical characteristics of its components, and not on the engineered structure. Due to the lack of long-term performance data, the PAs developed for DOE LLW disposal sites give only short-term credit to waste form performance in the algorithms used to determine exposure (DOE, 1998b).[6] This approach introduces a high degree of conservatism into PA exposure calculations by incorporating radionuclide release rates that are almost certainly greater than would actually occur in a disposal facility. This in turn leads to a higher calculated dose and a shorter travel time to persons exposed through the water pathway.

The IT Corporation (1993) performed a set of PA calculations for two generic MLLW disposal facilities, one located in a humid climate and the other in an arid region. This analysis showed the waste form can be expected to do little to reduce the annual dose that a critical group far in the future might receive from radionuclides with half-lives greater than 1,000 years. The principal factors that determine the integral radiological dose from long-lived nuclides are the total waste inventory and the site's hydrogeological and geochemical characteristics. The IT study found that sites in arid regions would release radionuclides at sufficiently low rates to be below DOE dose limits.

The conclusion that the leach rate of long-lived radionuclides from a waste form is less important than site hydrogeological conditions in limiting exposure was also reached in the performance evaluation study conducted by Waters, et al. (1996). This study investigated the technical feasibility of disposing grout-stabilized mixed waste (other

---

[6] The lack of long-term performance data for waste forms was confirmed by a literature search of relevant reports from waste form developers. The reports provided no quantitative longevity data that could serve as input to PA models.

waste forms were not considered) at 15 DOE sites throughout the country using a simplified PA process. The study found that waste disposal facilities in arid climates with low infiltration rates, large depths to ground water, and low ground water velocities could accept large amounts of MLLW without exceeding DOE dose limits. In contrast, waste facilities in humid climates with high infiltration rates, shallow depths to ground water, and high ground water velocities could only accept low amounts of long-lived radionuclides that are soluble in water (i.e., ^{99}Tc). The performance of the waste form had little effect on the analysis because it was assumed that extensive leaching would occur during the 10,000 year period covered in the analysis. Although the waste form was assumed to have decomposed by the time the integrity of the facility was lost, a sensitivity analysis performed by Waters, et al. (1996) showed that the retention of radionuclides by the residual components of the grouted waste form (i.e., by sorption) and the annual infiltration rate were the parameters that had the most significant effect on the dose from the water exposure pathway. Both studies (Waters, et al., 1996; IT Corporation, 1993) concluded that limiting the long-lived soluble radionuclides in waste disposal facilities in humid regions of the country may be necessary to reduce the long-term dose that might occur through the water exposure pathway.

It is worth noting that in spite of the high degree of conservatism in the analysis conducted by Waters, et al. (1996), the analysis showed that most of EM's MLLW waste could be disposed of at existing DOE sites. Another report (Waters, et al., 1998) considered 6,250 m^3 of treated MLLW identified as potentially problematic (i.e., containing radionuclides at concentrations potentially too high to permit disposal). This study determined that this problematic waste could in fact be disposed safely at arid sites.[7] The waste forms assumed in the study were based on treatment plans developed by individual DOE sites. In this context, about 95% of the waste was in grout form, about 4% was vitrified, and the remainder was macroencapsulated in polymer. These modeling studies tend to confirm that disposal options for EM's mixed waste are not

[7] About 1,800 m^3 of this waste could be disposed at the Envirocare MLLW facility in Utah, and all but 100 m^3 of the waste met waste acceptance criteria for the Hanford LLW disposal facility.

limited by a need for new or better waste forms. This conclusion was also reached in the Waste Form Initiative Close Out Report (DOE, 1998b).

Although RCRA does not require knowledge of the long-term performance of stabilized hazardous wastes, Franz, et al. (1994) applied a PA methodology to calculate releases of hazardous metals, including lead (Pb), chromium (Cr), and mercury (Hg) from a hypothetical generic underground disposal facility. The parameters used in the calculations were chosen to simulate releases from one generic facility in a humid environment and from one in an arid environment. The objective was to determine potential leaching criteria that might be appropriate for the release rate of RCRA constituents from a stabilized waste. The principal focus of the study was to evaluate the effect of a waste form's fractional release rate on the contaminant concentration at the boundary of the disposal unit. The fractional release rate is the fraction of the total mass of contaminants in the waste released in one year. Fractional release rates ranging from 10^{-8}/yr to 10^{-2}/yr were used in the calculations. A conservative modeling approach was used that neglected the effects of solubility on contaminant concentrations, and did not consider sorption reactions by soil constituents. This model is conservative in the sense that the neglected effects reduce contaminant concentrations at the boundary, thus, neglecting them gives a worst-case scenario. The results of this analysis were that a fractional release rate of 10^{-5}/yr is appropriate for RCRA metals as well as for long-lived radionuclides (Franz, et al., 1994).

This study is perhaps the first to establish a numerical leaching criteria based on the modeled performance of a disposal facility containing RCRA hazardous constituents. However, the fact that it is based on generic site parameters introduces considerable uncertainty to the resulting fractional release rate. Nevertheless, it does give an order-of-magnitude estimate of the performance that might be required for stabilized wastes containing both hazardous metals and long-lived radionuclides. The study helped confirm that the fractional release rate, a parameter that is typically measured in waste form characterization tests, is a key parameter for assessing long-term performance of waste forms that contain both hazardous and radioactive wastes.

AIRBORNE EXPOSURE SCENARIOS

There are only a few radionuclides in the DOE inventory that can exist in the vapor phase at normal temperatures, notably radon, carbon-14 as $^{14}CO_2$, and tritium in the form of tritiated water. While many RCRA wastes (such as solvents) are volatile, they are expected to be removed during waste treatment. If gases are present after treatment, their low viscosity will allow them to move freely through most waste forms considered for MLLW disposal. For these reasons, most PAs do not evaluate the ability of typical waste forms, such as those described in Chapter 4, to retain volatile constituents.

The USNRC (1997) has developed guidance on developing PA analysis for gaseous releases from LLW disposal facilities. They are predicated on the containment of gases in high integrity containers rather than in typical waste forms. Two release scenarios are proposed: (1) all the containers simultaneously fail, resulting in a puff release; and (2) the entire inventory of ^{14}C, ^{3}H, ^{85}Kr, ^{222}Rn, and ^{129}I in the disposal facility is available for release during the time period that is considered in the assessment (e.g., 1 year).

Waters, et al., 1996, conducted a performance evaluation (PE) study to consider site constraints for waste disposal at 15 DOE sites throughout the country. This PE considered only ^{3}H and ^{14}C. The analysis used the approach for the PA documents written for LLW disposal facilities at Hanford, Idaho National Engineering and Environmental Laboratory, Nevada Test Site, Oak Ridge Reservation, and Savannah River Site. Volatile radionuclides were assumed to be transported to the soil surface by diffusion in the vapor phase, and then transported and dispersed in the atmosphere according to an analytical Gaussian dispersion model. A grouted waste form was assumed in this analysis, and for conservatism, it was further assumed that its diffusive properties were similar to those of the native soil. Therefore, no credit was taken for the waste form or the disposal facility's ability to reduce emanation of the volatile constituents. Furthermore, 100 years of retention in the disposal facility were assumed prior to release of the radionuclides. This provided sufficient time for most of the ^{3}H (with a half-life of about 12.1 years) to decay. Based on the PE analysis, permissible amounts of ^{14}C to be disposed in the facility would be limited by the atmospheric pathway at eight of the 15 sites, but there would be no limit for ^{3}H because of its decay. The eight sites are located mainly in arid regions where low water infiltration rates reduce potential waterborne exposure scenarios.

One factor not considered in the PE analysis was the interaction between the $^{14}CO_2$ and the grout used to stabilize the waste. Aqueous leachates from cementitious grouts have high pH values that would increase the solubility of $^{14}CO_2$ in the liquid phase by orders of magnitude so that much less than the amount calculated in the PE could become airborne. Furthermore, portland cement-based grouts are hygroscopic and would likely scavenge much of the tritiated water preventing its release to the atmosphere. Waters, et al. (1996) noted that even if these retention mechanisms are ignored and hence the model is overly conservative, atmospheric exposure scenarios do not limit the permissible concentrations of tritium and ^{14}C for disposal of most LLW waste constituents.

INTRUSION SCENARIOS

Inadvertent intrusion scenarios are constructed hypothetically as a means to estimate the level of protection that the disposal system would provide if it should be breached in the future due to a loss of societal memory regarding the nature of the waste. Intrusion scenarios leading to exposure to radioactive wastes are notoriously difficult to quantify, because they depend on future activities under conditions that are not possible to foresee with any degree of certainty. Intrusion scenarios are generally assumed to be inadvertent since access to the waste disposal facility is unintentional and results from some other activity, such as construction, excavation, or drilling. The assumption of inadvertent intrusion into a waste disposal facility was used in developing the waste classification system for near-surface disposal of LLW contained in 10CFR61 (i.e., Class A, B, and C wastes), discussed in Chapter 3. Inadvertent intrusion scenarios assume that the integrity of the waste form is compromised by excavation or drilling activities, and therefore no credit for contaminant immobilization is claimed. Waste release is immediate and no attenuation occurs.

There are two important assumptions required to evaluate inadvertent intrusion scenarios: (1) the nature of the applicable scenarios, and (2) the time at which intrusion occurs. The applicable scenarios will depend primarily on the site conditions and facility design. There are at least five exposure pathways for MLLW: (1) inhalation of contaminated dust, (2) ingestion of contaminated soil, (3) ingestion of contaminated agricultural products, (4) dermal contact with contaminated soil, and (5) exposure to radiation from the radioactive waste materials. The possibility of burying waste deeper than a homesteader's expected excavation represents one possibility for minimizing intrusion. Placing the waste in a concrete vault or tumulus is another. The time at which inadvertent intrusion is assumed to occur will affect the remaining activity of short-lived radionuclides. DOE Order 5820.2A specifies that MLLW disposal facilities will be controlled for 100 years after closure. For the homesteader scenario, Waters, et al. (1996) assumed an inadvertent intrusion at 300 years following closure for a RCRA-compliant trench and 500 years for a concrete tumulus. An inadvertent intrusion at 100 years following closure was assumed for the post-closure drilling scenario. Due to its great depth, the inadvertent human intrusion scenarios for the Waste Isolation Pilot Plant (WIPP) are all predicated on future drilling of a borehole while exploring for mineral resources (Rechard, 1995). In reviewing progress on the WIPP, a National Research Council (NRC) committee observed that it would be unfortunate if such intrusion scenarios, made without a scientific basis, resulted in disqualification of the site (NRC, 1996b).

The role of the waste form in inadvertent intrusion scenarios has not been extensively considered to date since no credit is usually taken for it. Conventional scenarios rely more on the facility location and design to prevent future exposure than on the waste form itself. The PE analysis by Waters, et al. (1996) assumed that the waste form would be indistinguishable from the native soil at the time of inadvertent intrusion and therefore accidental post-closure intrusion scenarios took no credit for the waste form. While this may be reasonable for a grout stabilized waste form, it may not be likely for a vitrified waste form, which should be much more durable. In fact, one scenario has been proposed in which an intruder encounters MLLW waste that has been stabilized as vitrified glass beads. The beads are so attractive that they are mined and subsequently incorporated in jewelry, possibly resulting in large doses of gamma radiation to the wearers. As a specific example, Pohl, et al.

(1996) developed a method of estimating the exposure risk associated with this pathway.

Waters, et al. (1996) noted that inadvertent human intrusion scenarios for PA calculations at LLW disposal facilities tend to be the most restrictive of the standard exposure scenarios evaluated. This finding was confirmed in their PE analysis of potential mixed waste disposal at 15 DOE sites throughout the country. This is particularly important for long-term, chronic exposure since a homesteader may encounter the radioactive waste materials over a long period of time and through a variety of different mechanisms, including inhalation, ingestion, and external exposure. At many sites, particularly arid sites, the inadvertent intrusion scenarios provide the most restrictive permissible waste concentrations for most radionuclides, because airborne and waterborne pathways are not significant (Waters, et al., 1996).

FINDINGS, DISCUSSION, AND RECOMMENDATIONS

PA is required by the DOE and the USNRC to evaluate the safety of radioactive waste disposal systems for future generations and the environment. The committee undertook to determine the role of the waste form in PA. The committee found that current PA models do not take significant credit for the waste form's ability to reduce the release rate of hazardous and radioactive constituents. This is mainly because of the lack of quantitative long-term release data that can be used in PAs and results in a conservative perspective with respect to the release of contaminants from a prospective disposal facility. More realistic assessments may allow more effective use of the capacity of disposal facilities by allowing them to accept a larger inventory of radionuclides or hazardous wastes.

Committee findings include the following:

- EPA regulations that include only short-term performance of the waste form [e.g., passing the Toxicity Characteristic Leaching Procedure (TCLP)] and the USNRC requirement for PA to extend beyond the useful life of waste forms both de-emphasize the role of the waste form in limiting the long-term release of waste constituents.

- Radioactive waste inventories allowed in USNRC-licensed facilities are limited by PA calculation of exposures that result from future intrusions into disposal facilities (intrusion scenarios). These calculations generally do not take credit for the waste form as a barrier.

Current EPA regulations require that the waste form meet short-term prescriptive criteria, such as the TCLP. USNRC regulations require PA evaluations that extend well beyond the expected useful life of any waste form. Taken together, these regulations de-emphasize the role of the waste form in limiting the long-term release of waste constituents. Between these extremes, however, the waste form can be expected to play a very important role, including near total confinement of the most common intermediate-lived radionuclides ^{137}Cs and ^{90}Sr, and very gradual release of long-lived radionuclides (such as ^{99}Tc) and hazardous waste constituents.[8]

Performance assessment is part of the regulatory process. If a PA is to take credit for the waste form, the waste form's long-term performance must be described well enough that the PA can withstand regulatory review. As discussed in Chapter 5, current understanding of long-term performance of the waste form is not sufficient for extrapolations over thousands of years. The limited role of waste forms in PA is not necessarily a failure of waste form development, but rather that the present understanding of their long-term performance is inadequate. The committee believes that the credibility of performance assessments can be enhanced by better representation of the waste form's behavior in the disposal environment.

Because performance assessment is a general requirement for disposal of essentially all types of DOE waste, not only mixed waste, the committee addresses the following recommendations to the Office of Science and Technology (OST, EM-50).

---

[8] ^{137}Cs and ^{90}Sr radioactive have half-lives of approximately 30 years, and their decay will effectively remove them from a waste inventory in 300-500 years. For practical purposes, long-lived nuclides like ^{99}Tc and many hazardous waste constituents, such as the heavy metals, will never decay.

- OST should support efforts to obtain data that will allow a more realistic inclusion of waste forms in PA models, including intrusion scenarios. Without such data the waste form will never receive proper credit in PA with the resulting cost penalties for additional engineered barriers and possible restriction in site selection.
- OST should play a more significant role in promoting (funding) cooperation among investigators who are characterizing waste forms and those who are developing PA models. This will help ensure that characterization data are useful for PA models, and that PA models properly incorporate this data.

7

Findings and Recommendations

The Committee on Mixed Wastes was requested by the Office of Science and Technology (OST, EM-50) to review and evaluate the state of development of the final waste forms of treated mixed wastes as they arise from current and emerging treatment technologies. The committee was also asked to identify the options the Department of Energy (DOE) might consider in technology development in order to achieve the desired waste forms that are cost-effective and safe for disposal. In carrying out the review, the committee received formal presentations from DOE staff and other individuals, and examined documents and data provided by DOE and from other sources. Some committee members visited waste contractors and DOE sites.

In the committee's view, the successful operation of any technology-based system is the result of system design and management. Discrepancies between desired and actual performance provide the impetus for research and development (R&D) in both technology and management sciences. This concept is a continuum of efforts aimed toward a defined goal, namely matching the actual system performance to the desired performance. In practice, the desired performance is defined by legislation, federal regulations, DOE cleanup needs, and stakeholder values.

The DOE Office of Environmental Management's (EM's) responsibility for cleaning up the weapons complex and disposing of the wastes makes the current mixed waste focus area (MWFA) program similar to an industrial R&D program focused on a fixed end point. Such R&D programs undergo a natural evolution of changing emphasis and scope. In the first stages, the R&D program is called on to define basic science and technology. As the program progresses, the role of R&D changes, shifting from a research mode to a developmental and demonstration mode and into a support mode. Having defined the technology, the R&D organization must then focus on supporting deployment of the technology. The committee views the role of OST and its MWFA as provider of scientific and technical support in all phases of this evolution.

MWFA presentations to the committee described a clear strategy for resolving technology deficiencies that fits within the scope of a supporting R&D organization such as OST.[1] However, the presentations indicated that MWFA is using only part of a comprehensive systems approach to waste management, which begins with the untreated waste and ends with closure of the disposal facility, and evolves stepwise from basic R&D to technical support for process deployment. The committee's recommendations discussed below reflect its view of the importance of the systems approach for technology development and deployment activities by the MWFA.

GENERAL FINDINGS AND RECOMMENDATIONS

The committee's primary recommendation is that MWFA should no longer emphasize the development of new classes of waste forms. After reviewing the technologies available to treat EM's mixed waste inventory, and considering the resulting waste forms, it is the committee's judgment that no new classes of waste forms are required. Clearly no single form is appropriate for all wastes, but collectively the variety of available waste forms and well-established waste form production technologies make it unlikely that any totally new class of waste forms will be necessary to complete EM's planned cleanup program. MWFA should now emphasize engineering design, integration, and scale-up of its proposed treatment processes and their demonstration and deployment at the DOE sites. Technology development and deployment

[1] Presentations heard by the committee are described in Appendix B.

must be cost-effective and commensurate with EM waste management strategy described in "Paths to Closure" (DOE, 1998c).

Secondly, the committee recommends that MWFA should continue its practice of identifying, prioritizing, and responding to technology deficiencies. As discussed in Chapter 4, the MWFA has established a rational and systematic program that identifies and prioritizes deficiencies. The committee compliments MWFA on this effort and encourages continued updates of the Technical Baseline Report to document the state of its technology development and deployment activities.

Thirdly, the committee recommends that MWFA should broaden its use of the systems approach, in selecting, developing, and deploying technologies. Such an approach would begin with characterization of the waste (development of cost-effective and efficient methods) and definition of the required performance of a proposed treatment technology, based on EM's needs, regulatory requirements, and stakeholder expectations. Using its technical and managerial resources, MWFA should develop and assist in the design, development, and deployment of the new technology to ensure that the technology meets its performance goals. An important aspect of a good systems approach is flexibility to accommodate new information experiences and reasonable changes in the desired output. Inevitably there are iterations between the actual and the desired performance of any waste management system as its technology matures and expectations change.

In recommending that MWFA broaden its systems approach, the committee recognizes that the technical issues pertaining to waste management are often overshadowed by non-technical (e.g., political and social) issues. Public acceptance of a waste management strategy may be transient, which creates a moving target for engineers and program planners. An example of these conflicts can be seen in the present study of currently available waste forms. The MWFA is developing and proposing technological methods to convert EM's mixed inventory into stable waste forms for disposal on a time schedule to meet the EM's "Paths to Closure" commitments. However, many of the disposal sites that will receive mixed waste forms, waste acceptance criteria, and the actual disposal conditions for the waste forms are unspecified. A comprehensive systems approach must recognize these complicating, but real, factors.

WASTE CHARACTERIZATION

The committee found that EM's mixed waste inventory is sufficiently well characterized for conceptual design of treatment processes but insufficiently characterized for detailed engineering design or process optimization. Detailed characterization, using currently available methods, will be expensive and may entail risks to operating personnel.

The Committee has three recommendations:

1. The MWFA should develop simplified methods to characterize the waste inventory, with emphasis on nondestructive examination and nondestructive assay techniques. Emphasis should be placed on developing better methods to determine heavy metals and solvent contamination.
2. The MWFA should continue to develop, demonstrate and encourage deployment of techniques and procedures to ensure that all new waste streams are adequately characterized.
3. The MWFA should strive for a balance between the risks, benefits, and cost of detailed characterization and the effort and cost to develop more robust treatment technologies that can handle a wide variety of waste compositions, thus reducing the required degree of waste characterization.

There are two possible pathways to reduce costs and risks for which R&D would be valuable. The first is for simplified methods of examining and characterizing wastes through such vehicles as rapid scanning, non-intrusive identification of constituents. The other pathway is through development of robust treatment technologies that can adequately deal with a variety of feed materials, thereby reducing the need for extensive characterization. The committee noted that the MWFA's recent experience in developing "universal" treatment technologies, such as the plasma torch, showed that these technologies still require adjustment according to waste composition. The committee therefore emphasizes the importance of waste characterization and encourages the MWFA to seek a reasonable balance in the development of characterization and treatment technologies.

The committee recognizes the difficulties in determining the composition of mixed wastes when the waste is stored in many forms or in sealed containers. Detection and determination of some of the radioactive constituents by non-intrusive means is readily accomplished. Similar detection and determination of 'silent' hazardous materials such as toxic organic compounds and metals or characteristic waste components that could pose processing problems during the generation of acceptable waste forms continues to represent a major theoretical and practical challenge. The committee believes research efforts devoted to this problem could pay significant dividends, particularly in reducing the quantity of waste that must be treated as mixed waste that is, waste minimization. However, the committee also believes that the development of robust and comprehensive processing techniques that are insensitive to waste composition and yield satisfactory waste forms when the feed to the process is only poorly characterized should also be pursued.

TREATMENT TECHNOLOGIES

In its Technical Baseline Report, MWFA has drawn up process diagrams that conceptualize treatment for the majority of its varied waste streams. This effort has defined potential treatment systems and their final waste forms to meet the present regulations governing mixed wastes. In addition, volume reduction, cost reduction, and suitability for transportation have also been objectives of the work.

The committee found that there are ample treatment technologies and waste forms for EM's mixed waste inventory. However, many of these technologies have not been demonstrated as part of an integrated production-scale system using actual wastes.

The Committee has four recommendations:

1. MWFA should integrate treatment technologies for its five treatment groups into a mixed waste treatment strategy. This strategy should consider the waste form as a part of an overall mixed waste management system that includes the following:

- compatibility of waste form with transportation and disposal options,
- trade-offs between risks to personnel associated with additional waste characterization and additional costs of a more robust treatment and stabilization system, and
- trade-offs between the increased number of disposal options for a very stable waste form, versus the lower costs but reduced disposal options for less stable waste forms.

2. MWFA should demonstrate new treatment technologies on at least the pilot plant scale using real wastes or realistic surrogates before the technology is designated as ready for deployment.

3. MWFA should continue to address technology deficiencies that it has identified through input from the Site Technology Coordinating Groups and update its Technical Baseline Report to reflect progress in addressing these deficiencies.

4. MWFA should continue to provide research funding for developing robust processes that can treat and stabilize waste of poorly defined or variable composition, recognizing the trade-off between better waste characterization and development of improved treatment technology.

The need for a treatment strategy follows from the committee's overall recommendation that MWFA adopt a more complete systems approach to its technology development, and from the committee's recognition of the trade-offs between the difficulties of developing broadly targeted, robust treatment technologies, and the potential risks to operators in sampling and analyzing the many and varied wastes that comprise the inventory. These were discussed in Chapter 2 and in the preceding recommendations on waste characterization. Trade-offs between grout and higher quality vitreous forms that are generally harder to make, as discussed in Chapter 4, are also recognized by the committee. The committee noted that the present plan for the privatized Advanced Mixed Waste Treatment Project is to produce grout forms, whereas early presentations by MWFA indicated that glass waste forms would also be produced.

The necessary technology demonstrations that are being planned at several DOE sites will inevitably be constrained by time and budgets. Such constraints can set the stage for later technology failures if larger scale testing is not done with careful planning, care, and diligence. For example, process steps should be thoroughly tested and evaluated before any radioactive materials enter a new facility. The final products should be extensively analyzed to avoid such issues as inadequate waste form performance because of unexpected variations in feedstock composition. The MWFA should play an important role as a technology advisor during this very important demonstration phase. MWFA should also remain closely involved in the technology deployment phase to ensure successful technology transfer to the DOE sites or private contractors.

WASTE FORM CHARACTERIZATION AND PERFORMANCE ASSESSMENT

To determine if current waste forms are sufficiently developed to stabilize EM's inventory of mixed waste, test protocols to characterize the waste forms must be available. The committee found no established tests that can demonstrate the long-term (greater than a few hundred years) behavior of a waste form; however, present methods are adequate to evaluate short-term behavior.

Performance assessment (PA) is required by the U.S. Nuclear Regulatory Commission (USNRC) to evaluate the long-term safety of waste disposal facilities. The committee found that current PA methodology does not recognize the significance of the waste form or take reasonable credit for the waste form's ability to reduce the release rate of hazardous and radioactive constituents. This is mainly because of the lack of methods to quantify the long-term behavior of waste forms.

The committee's four recommendations in the areas of characterization and performance assessment are directed to OST because the recommendations apply to all DOE wastes:

1. OST should continue to support programs aimed at fundamental understanding of waste form durability and degradation processes. These programs should lead to a better representation of the waste form in PA modeling.
2. OST should work to promote consensus among the U.S. Environmental Protection Agency (EPA), USNRC, DOE, and the scientific community on waste form testing methods that will be generally acceptable for providing at least a qualitative evaluation of long-term waste form performance in disposal environments.
3. OST should support efforts to obtain data that will allow a more realistic inclusion of waste forms in PA models, including inadvertent intrusion scenarios. Without such data the waste form will never receive proper credit in PA with the resulting cost penalties for additional engineered barriers and possible restriction in site selection.
4. OST should play a more significant role in promoting (funding) cooperation among investigators who are characterizing waste forms and those who are developing PA models to help ensure that characterization data are useful for PA models, and that PA models properly incorporate this data.

The credibility of performance assessments can be enhanced by better representation of waste form behavior in the disposal environment. More realistic assessments might allow more effective use of the capacity of disposal facilities by allowing them to accept a larger inventory of radionuclides or hazardous wastes. The EM Science Program (NRC, 1997) could provide a valuable mechanism for evaluating and funding research proposals for the fundamental studies recommended by the committee.

REGULATORY GUIDELINES

The lack of available, licensed disposal sites for mixed waste and uncertainties in the waste acceptance criteria of future sites introduce a significant risk in judging the adequacy of EM's planned mixed waste treatments and waste forms. The committee recommends that EM work with EPA and the USNRC to agree on clear guidelines that describe acceptable waste forms for disposal of mixed waste in future, near-surface disposal facilities. This should be done as soon as possible to reduce the risk of EM deploying technologies that are later judged inadequate because of unanticipated regulatory requirements.

References

Amendola, A., S. Contini, and I. Zromas. 1992. Uncertainties in Chemical Risk Assessment: Results of a European Benchmark Exercise. Journal of Hazardous Materials. February. 29(3): 347.

American Nuclear Society (ANS). 1986. American National Standards Measurement of the Leachability of Solidified Low-Level Radioactive Wastes by a Short-Term Test Procedure. ANS Standards Committee Working Group 16-1. ANSI/ANS-16.1-1986.

Atkins, M., D. Damidot, and F. P. Glasser. 1994. Performance of Cementitious Systems in the Repository. In *Scientific Basis for Nuclear Waste Management XVII*. Vol. 333. A. Barkatt and R. A. Van Konynenburg, (Eds.). Pp. 315-326. Pittsburgh, Pa.: Materials Research Society.

Berry, J. B. 1994. Mixed Waste Integrated Program Strategic Plan, Revision B (draft). 1994. Office of Technology Development. U.S. Department of Energy.

Case, D. R. 1991. Resource Conservation and Recovery Act. In *Environmental Law Handbook*, 11th ed. J. G. Arbuckle et al. (Eds.). Pp. 406-442. Rockville, Md.: Government Institutes, Inc.

Conner, J. R. 1990. *Chemical Fixation and Solidification of Hazardous Wastes*. New York: Van Nostrand Reinhold.

Connolly, M. 1997. Review of the Current Status of Mixed Waste Focus Area Efforts. Idaho Falls, Id. September 15.

Cunnane, J. C. et al. (Eds.). 1994. High-Level Waste Borosilicate Glass: A Compendium of Corrosion Characteristics. Vols. I, II, and III. DOE/EM 0177. Office of Waste Management. U.S. Department of Energy.

Donald, I. W., B. L. Metcalfe, and R. N. J. Taylor. 1997. Review: The Immobilization of High Level Radioactive Wastes Using Ceramics and Glasses. Journal of Materials Science. 32: 5851–5887.

Electric Power Research Institute (EPRI). 1997a. Advanced Mixed Waste Treatment Technologies: Results of Mixed Waste Treatment at the M-4 Facility. EPRI-TR-107974. September.

Electric Power Research Institute. 1997b. Mixed Waste Advanced Treatment Technologies: Waste Processing Products and Their Recycling Application. EPRI-107990. October.

Evans, L., A. Pernstein, and E. Wong, 1997. Application of Quantum-CEP Technology to Utility Mixed Waste. Journal of the Franklin Institute. 334A.

Ewing, R. C., and M. J. Jercinovic. 1987. Natural Analogues: Their Application to the Prediction of the Long-Term Behavior of Nuclear Waste Forms. In *Scientific Basis for Nuclear Waste Management*. X. J. K. Bates and W. B. Seefeldt (Eds.). Materials Research Society Proceedings. 84: 67–86. Pittsburgh, Pa.

Ewing, R. C., W. I. Weber, and F. W. Clinard, Jr. 1995. Radiation Effects in Nuclear Waste Forms. Progress in Nuclear Energy: An International Review Journal. 29(2): 63–127.

Federal Facility Compliance Act (FFCA). 1992. Public Law 102-386. Available from the FFCA Bulletin Board. Internet http://www.em.doe.gov/ostp/#ba.

Franz, E. M., M. Fuhrmann, B. Bowerman, S. Bates, and R. Peters. 1994. Proposed Waste Form Performance Criteria and Testing Methods for Low-Level Mixed Waste. BNL-52436. Upton, N.Y.: Brookhaven National Laboratory.

Fuhrmann, M., R. Pietrzak, J. H. Heiser, E. M. Franz, and P. Colombo. 1990. Method for Accelerated Leaching of Solidified Waste. BNL-52268. Upton, N.Y.: Brookhaven National Laboratory.

Gershey, E. L., R. C. Klein, E. Party, and A. Wilkerson. 1990. *Low-Level Radioactive Waste: From Cradle to Grave.* New York: Van Nostrand Reinhold.

Gilliam, T. M., and C. Wiles (Eds.). 1996. *Stabilization and Solidification of Hazardous, Radioactive, and Mixed Wastes.* West Conshohocken, Pa.: American Society for Testing and Materials.

International Atomic Energy Agency (IAEA). 1996. Clearance Levels for Radionuclides in Solid Materials: Application of Exemption Principles. Interim Report for Comment. IAEA-TECDOC-855.

IT Corporation. 1993. Mixed Low-Level Waste System Analysis Methodology and Applications Report (draft). Vol. 1. Radioactive Waste Technical Support Program. DOE/LLW-194.

Jantzen, C. M., N. E. Bibler, D. C. Beam, W. G. Ramsey, and B. J. Waters. 1993. Standard Test for Determining Chemical Durability of Nuclear Waste Glasses: The Product Consistency Test (PCT). ver. 6.0. Submitted to ASTM C-26.13.

Kalb, P., D. J. R. Conner, J. L. Mayberry, B. R. Patel, J. M Perez, and R. L. Treat. 1997. *Stabilization/Solidification.* Annapolis, Md.: American Academy of Environmental Engineers.

Kirk, D. R. 1996. Summary of USEPA Research on Solidified Waste for Long-Term Durability. In *Stabilization and Solidification of Hazardous, Radioactive, and Mixed Wastes.* Vol. 3. T. M. Gilliam and C. C. Wiles (Eds.). ASTM STP 1240. West Conshohocken, Pa.: American Society for Testing and Materials.

Kolts, J. 1996. Presentation to the NRC Mixed Waste Subcommittee. Overview of the Mixed Waste Problem. Idaho Falls, Id. November 19-20.

Lee, J. H., D. M. Roy, B. Mann, and D. Stahl. 1995. Integrated Approach to Modeling Long-Term Durability of Concrete Engineered Barrier in LLRW Disposal Facility. In *Scientific Basis for Nuclear Waste Management XVIII.* Vol. 353. R. C. Ewing and J. Murakami (Eds.). Pp. 881–889. Pittsburgh, Pa.: Materials Research Society.

Lutze, W., and R. C. Ewing (Eds.). 1988. *Radioactive Waste Forms for the Future*. Amsterdam: North-Holland Physics Publishing.

Mayberry, J. L., and T. L. Huebner. 1993. Technical Area Status Report for Low-Level Mixed Waste Final Waste Forms. DOE/MWIP-3. Idaho Falls, Id.: Science Applications International Corporation.

Means, J. L., L. A. Smith, K. W. Nehring, S. E. Brauning, A. R. Gavaskar, B. M. Sass, C. C. Wiles, and C. I. Mashni. 1995. *The Application of Solidification/Stabilization to Waste Materials*. Boca Raton, Fla.: Lewis Publishing.

National Research Council (NRC). 1996a. Glass as a Waste Form and Vitrification Technology: Summary of an International Workshop. Washington, D.C.: National Academy Press.

National Research Council. 1996b. The Waste Isolation Pilot Plant: A Potential Solution for the Disposal of Transuranic Waste. Washington, D.C.: National Academy Press.

National Research Council. 1996c. Environmental Management Technology Development Program at the Department of Energy, 1995 Review. Washington, D.C.: National Academy Press.

National Research Council. 1997. Building an Effective Environmental Management Science Program: Final Assessment. Washington, D.C.: National Academy Press.

Perret, D. 1998. Municipal Solid Waste (MSW) Glasses: The Opinion of Selected Experts. Report 2. Swiss Agency for Environment, Forests and Landscape. October.

Pohl, P. I., W.-C. Cheng, T. Wheeler, and R. D. Waters. 1996. Waste Form Performance Assessment Task. Letter Report. TTP AL2-6-TI-01. Albuquerque, N. Mex.: Sandia National Laboratories.

Poon, C. S. 1989. A Critical Review of Evaluation Procedures for Stabilization/Solidification Processes. Environmental Aspects of Stabilization and Solidification of Hazardous and Radioactive Wastes. P. Cote and M. Gilliam (Eds.). ASTM STP 1033. Philadelphia, Pa.

Quillin, K. C., S. L. Duerden, and A. J. Majumdar. 1994. Accelerated Aging of Blended OPC Cements. In *Scientific Basis for Nuclear Waste Management XVII*. Vol. 333. A. Barkatt and R. A. Van Konynenburg (Eds.). Pp. 341–347. Pittsburgh, Pa.: Materials Research Society.

Rechard, R. P. 1995. An Introduction to the Mechanics of Performance Assessment Using Examples of Calculations Done for the Waste Isolation Pilot Plant Between 1990 and 1992. SAND 93-1378. Albuquerque, N. Mex.: Sandia National Laboratories.

Schwinkendorf, W. E., B. C. Musgrave, and R. N. Drake. 1997. Evaluation of Alternative Nonflame Technologies for Destruction of Hazardous Organic Waste. INEL/EXT-97-00123. April.

Seitz, R. Assessment of Health and Environmental Effects from Radioactive and Chemically Toxic Waste. In *Proceedings from the International Workshop on Comparative Evaluation of Health Effects of Environmental Toxicants from Advanced Technologies, January, 28-30, 1998, Chiba City, Japan.*

Sewards, T., et al. 1991. Mineralogy of the Rustler Formation in the WIPP-19 Core. SAND87-7036. Albuquerque, N. Mex.: Sandia National Laboratories.

Shevenell, L., and F. O. Hoffman. 1993. Necessity of Uncertainty Analysis in Risk Assessment. Journal of Hazardous Materials. December. 35(3): 369.

Singh, D., D. Barber, A. S. Wagh, R. V. Strain, and M. Tlustochowicz. 1998. Stabilization and Disposal of Argonne-West Low-Level Mixed Wastes in Ceramicrete Waste Forms. Idaho Falls, Id.: U.S. Department of Energy.

Sullivan, T. F. P. (Ed.). 1997. *Environmental Law Handbook.* 14th ed. Rockville, Md.: Government Institutes, Inc.

Sullivan, T. N. 1993. Disposal Unit Source Term (DUST) Data Input Guide. BNL-52375. Upton, N.Y.: Brookhaven National Laboratory.

Swedish Environmental Protection Agency (SEPA). 1998. Report 4752. Stockholm, Sweden.

U.S. Department of Energy (DOE). 1995a. The DOE National 1995 Mixed Waste Inventory Report. Internet http://www.em.doe.gov/ffcabb/ffcaftp/index.html. January 1999.

U.S. Department of Energy. 1995b. Index of FFCA Act Bulletin Board Downloadable Documents and Files. Internet http://www.em.doe.gov/ffcabb/ffcaftp/index.html. June 1997.

U.S. Department of Energy. 1996a. Mixed Waste Focus Area Integrated Technical Baseline Report. DOE/ID-10524. Vols. 1 and 2. Idaho Falls, Id. January.

U.S. Department of Energy. 1996c. Interim Format and Content Guide and Standard Review Plan for U.S. Department of Energy Low-Level Waste Disposal Facility Performance Assessments. Washington, D.C. October.

U.S. Department of Energy. 1997a. Mixed Waste Focus Area Technical Baseline Report. DOE/ID-10524. Vols. 1 and 2. Idaho Falls, Id. April.

U.S. Department of Energy. 1997b. Mixed Waste Focus Area Annual Report. Washington, D.C. April.

U.S. Department of Energy. 1998a. Order DOE 5400.1. General Environmental Protection Program. November 8, 1988, Change 1. June 29, 1990. Washington, D.C.

U.S. Department of Energy. 1998b. Waste Form Initiative Close Out Report (draft). INEEL EXT-98-00105. Idaho Falls, Id. February.

U.S. Department of Energy. 1998c. Accelerating Cleanup: Paths to Closure. DOE/EM-0362. Washington, D.C. June.

U.S. Department of Energy. 1998d. Deactivation and Decommissioning Focus Area. Quarterly Report. Washington, D.C. April.

U.S. Environmental Protection Agency (EPA). 1987. Joint USNRC-EPA Guidance Conceptual Design Approach for Commercial Mixed Low-Level Radioactive and Hazardous Waste Disposal Facilities. OSWER Directive #9487.00-8. 1987.

U.S. Environmental Protection Agency. 1996. United States Code of Federal Regulations, Title 40. Washington, D.C.: U.S. Government Printing Office. July 1.

U.S. Environmental Protection Agency. 1999a. Approach to Reinventing Regulations on Storing Mixed Low-Level Radioactive Waste: Advance Notice of Proposed Rulemaking, Federal Register. 64(39): 10063–10073. March 1.

U.S. Environmental Protection Agency. 1999b. Draft Technical Guidance for RCRA/CERCLA Final Covers.

U. S. Nuclear Regulatory Commission (USNRC). 1997. Branch Technical Position on a Performance Assessment Methodology for Low-Level Radioactive Waste Disposal Facilities (draft). NUREG-1573. Washington, D.C. May.

U. S. Nuclear Regulatory Commission. 1996. Licensing Requirements for Land Disposal of Radioactive Waste. Code of Federal Regulations. Title 10 Part 61. 1996 edition.

Wagh, A. S., S. Y. Jeong, D. Singh, R. Strain, H. No, and J. Wescott. 1997. Stabilization of Contaminated Soil and Wastewater with Chemically Bonded Phosphate Ceramics. Argonne, Ill.: U.S. Department of Energy.

Waters, R. D., et al. 1996. Performance Evaluation of the Technical Capabilities of DOE Sites for Disposal of Mixed Low-Level Waste. DOE/ID-10521. Albuquerque, N. Mex.: Sandia National Laboratories. (For a more current version of this report, see Waters, R. D., and M. M. Gruebel. 1999a. Performance Analysis for Disposal of Mixed Low-Level Waste I: Methodology and II: Results, Journal Environmental Engineering. 125: 249–266.)

Waters, R. D., P. I. Pohl, W. C. Cheng, M. M. Gruebel, T. A. Wheeler, and B. S. Langkopf. 1998. Evaluation of the Capabilities of the Hanford Reservation and Envirocare of Utah for Disposal of Potentially Problematic Mixed Low-Level Waste Streams. SAND98-0450. UC-2020. Albuquerque, N. Mex.: Sandia National Laboratories.

Weber, W. J., et al. 1997. Radiation effects in glasses used for immobilization of high-level waste and plutonium disposition. Journal of Materials Research. 12(8): 1946–1978.

Weber, W. J., et al. 1998. Radiation effects in crystalline ceramics for the immobilization of high-level nuclear waste and plutonium. Journal of Materials Research. 13(6): 1434–1494.

Wentz, C. A. 1989. *Hazardous Waste Management.* New York: McGraw- Hill, Inc.

APPENDIX A

Statement of Task

The Mixed Waste Committee will review and evaluate the state of development of the final waste forms of treated mixed wastes as they arise from current and emerging treatment technologies. In particular, the study will assess the characteristics of, and uncertainties associated with, the different types of final mixed waste forms for disposal, and identify the requirements for additional R&D. The study also will identify the options DOE should consider in technology development in order to achieve the desired mixed waste forms that are cost-effective and safe for disposal. The study will be based on information on waste to be treated, treatment technologies, waste form characterization studies, and anticipated disposal conditions.

APPENDIX B

Information Used by the Committee During Its Review

A) Presentations to the committee by various groups, especially by representatives of the Mixed Waste Focus Area (MWFA) of the DOE Office of Science and Technology

Overview of the Mixed Waste Problem, November 19-20, 1996, John Kolts, DOE–Idaho.

An overview of the mixed waste problem, including inventory, mission, MWFA organization, MWFA activities, technology deficiencies, and technology development requirements. (DOE, 1995a).

Mixed Waste Focus Area, Waste Form Initiative, November 16, 1996, Ron Nakaoka, DOE–Idaho.

A description of the Mixed Waste Focus Area, including background, the MWFA Waste Form Initiative, status of MLLW disposal facilities, Waste Acceptance Criteria for existing MLLW disposal facilities, and issues associated with MLLW disposal. The presentation described current technology deficiencies, the performance evaluation by Sandia National Laboratories, and the status of MLLW disposal sites. (DOE, 1996a).

Treatment and Stabilization of Mixed Wastes, Waste Form Testing and Evaluation, March 13, 1997, Ian L. Pegg, Vitreous State Laboratory, The Catholic University of America, Washington, D.C.

A discussion of waste form types, evaluation criteria, EPA hazardous waste regulations and tests, waste form performance and standard test methods, long-term performance testing of a glass waste from Savannah River, and a discussion of leaching mechanisms.

Mixed Waste Focus Area Waste Form Initiative–Technical Basis, Strategy and Plans, March 13, 1997, Jenya Macheret, DOE Idaho.

A review and recapitulation of the current status of Mixed Waste Focus Area activities.

Resource Conservation and Recovery Act: Overview and Mixed Waste Issues, March 13, 1997, Denise Glore, DOE–Idaho.

An overview of the EPA hazardous waste regulations as they affect the treatment and disposal of mixed waste.

Contamination Attenuation Mechanisms: Dispersion, Sorption, and Degradation, March 13, 1997, Patrick V. Brady, Sandia National Laboratories.

A discussion of the mechanisms by which contaminants are naturally attenuated in soils and ground water.

Results of Scoping-Level Analysis on Disposal of Treated LLW, March 13, 1997, Robert D. Waters, Sandia National Laboratories.

The approach used and the result of a performance evaluation of 15 sites. (Waters, 1996).

Current Management of Non-DOE LLMW, June 28, 1997, Carey A. Johnson, EPA Office of Radiation and Indoor Air.

An overview of mixed waste problems outside the scope of DOE activities, such as research facilities (both medical and non-medical), pharmaceutical research, medical applications, industrial applications, nuclear power plants, DOD, Superfund and RCRA cleanups. The presentation included a discussion of facilities that accept small quantities of radioactive materials (e.g., radioactive medical waste) for preprocessing and volume reduction

Mixed Waste Focus Area Waste Form Initiative, September 15, 1997, Ron Nakaoka, DOE–Idaho.

A review of the status of the waste form initiative with emphasis on site follow-up to the conclusions regarding the technical capability of individual sites to dispose of the majority of the mixed waste. (DOE, 1997a).

Mixed Waste Focus Area, September 15, 1997, Mike Connolly.

A review of the current status of mixed waste focus area efforts, including revisions to the prioritized list of technology deficiencies. Brief discussions of programs designed to address deficiencies such as the high temperature melter, characterization, TRU transportation, continuous emission monitors, mercury contamination, alternative organic oxidation, and salt and ash stabilization (DOE, 1997a).

Status of DOE Privatization, September 15, 1997, Jan M. Chavez, DOE–Idaho.

A summary of budget and funding requests for major privatization efforts, including the Advanced Mixed Waste Treatment Project.

Idaho National Engineering and Environmental Laboratory, Advanced Waste Treatment Project, September 15, 1997, BNFL, Inc.

An overview of a project to construct and operate a $1.16 billion facility at the Idaho site to treat approximately 25,000 cubic meters of waste.

In addition to the presentations, much information and data were obtained in discussions with the presenters and in response to questions.

B) Review of key DOE documentation

The baseline Mixed Waste Inventory Report–1995 referred to by OST was examined. At the time of compilation, the database contained the most up-to-date information available concerning the DOE inventory of waste materials, including quantity, hazardous and radioactive constituents, and proposed grouping for treatment and evaluation of data completeness and integrity (DOE, 1995a).

The Mixed Waste Focus Area has continued to revise and update inventory information. The latest edition of its report, "Mixed Waste Focus Area Technical Baseline Report," (DOE, 1997a) better describes expected treatment scenarios. A new prioritized list of technology deficiencies is presented. This report, in concert with the earlier version and the initial inventory, has been used as the basis for conclusions concerning characterization, needs for technology development, and applied engineering.

In the report "Performance Evaluation of the Technical Capabilities of DOE Sites for Disposal of Mixed Low-Level Waste," (Waters, et al., 1996), Sandia National Laboratories evaluates the capability of DOE sites to dispose of mixed waste materials.

C) Review of present disposal options

A commercial disposal site, Envirocare of Utah was visited to understand how site parameters, site permits, and waste acceptance criteria affect the receipt of waste materials. The WIPP site was visited and WIPP waste acceptance criteria were studied in similar fashion.

APPENDIX C

Technology Needs Identified by the MWFA

Technology Need	Summary Description
1. Characterization	Nondestructive examination and nondestructive assay techniques and equipment are required to determine the nature of the waste matrix in any package, confirm the presence and concentration of regulated materials and radionuclides, and identify characteristics of concern for operational safety and process continuity.
2. Mercury Stabilization	Mercury contamination at a level less than 260 ppm requires stabilization to control mercury solubility to the Universal Treatment Standards (<0.2 ppm).
3. Salt Stabilization	Stabilization processes are required that increase salt waste loadings, improve durability and/or reduce the volume increase typical of today's standard practices.
4. Waste Form Performance	An objective, technically defensible evaluation of the value of advanced waste forms in disposal site performance as-

	sessments is being conducted. Data needs identified in the evaluation will be addressed as required.
5. Material Handling	Methods and equipment designs are required that will provide for handling all types of DOE waste materials in all process steps without undue risk of exposure of operating personnel to radioactivity.
6. Sorting/Segregation	Efficient separation of mixed wastes from nonradioactive waste or waste that is radioactive only (i.e., non-mixed waste), in a manner that is safe, reliable, and minimizes exposure, is required.
7. Mercury Separation/ Removal	New techniques must be developed to physically or chemically remove the mercury from waste matrices (including soil, all types of process residues or sludges and particulate materials, and debris) for separate stabilization.
8. Mercury Amalgamation	Methods and equipment designs are required for amalgamating bulk non-recyclable mercury to meet the Universal Treatment Standards (<0.2 ppm).
9. TRU Transportation	Improved methods would increase the amount of untreated waste shipped to WIPP or a treatment facility, thereby reducing risks and costs associated with repackaging and/or treatment. Improved methods would reduce gas generation potential, reduce flammable gas concentrations, or provide alternative approaches to demonstrating compliance with allowable gas generation rates.

10. Ash Stabilization	Ash stabilization processes are required that increase waste loadings, improve durability, and/or reduce the volume increase typical of today's standard practices.
11. Mercury Monitor	Though mercury monitors are commercially available, it would be advantageous to develop real-time continuous emission monitors requiring minimal consumables and low maintenance, with operating ranges covering the emission limits typical of thermal treatment processes.
12. Alpha Monitor	Though alpha monitors are commercially available, it would be advantageous to develop real-time continuous emission monitors requiring minimal consumables and low maintenance, with operating ranges covering the emission limits typical of alpha material processing facilities.
13. Mercury Filter	A potential enhancement to traditional treatment design for selective mercury removal from off-gas, which removes essentially all of the mercury from the off-gas for separate treatment, is required.
14. Heavy Metal Monitor	It would be advantageous to develop real-time, multi-metal continuous emission monitors requiring minimal consumables and low maintenance, which can identify specific metals in operating ranges covering the emission limits typical of hazardous waste incinerators.
15. VOC/SVOC Monitor	It would be advantageous to develop real-time, multi-metal continuous emission monitors requiring minimal consumables and low maintenance, which can identify specific volatile and semi-volatile organic

	contaminants (VOC/SVOC), particularly dioxins and furans in operating ranges covering the emission limits typical of hazardous waste incinerators.
16. Alternative Organic Oxidation	Candidate technologies that are alternatives to incineration for oxidation of organics need to be demonstrated to verify whether any one or combination of technologies can reliably treat all of the organic constituents expected to be present in mixed waste, and be operated in a radioactive environment.
17. High Temperature Particulate Removal	Filters capable of operating at high temperatures, typical at the outlet of thermal treatment processes, and capable of removing a substantial fraction of the particulates prior to quenching or scrubbing are required.
18. Radionuclide Partitioning	More complete information on the partitioning of radionuclides between the final waste form, the off-gas, and any secondary wastes in mixed waste treatment processes is needed to support equipment design and process permitting.
19. Trace Metal Removal	Techniques are needed to meet permit requirements in effluents (e.g., 0.001 mg/L cadmium, 0.003 mg/L lead, and 0.004 mg/L silver) while minimizing secondary waste generation.
20. Fission Product Removal	Methods are needed for removal or significant reduction of the concentrations of fission products from mixed waste, especially process residues, sludges and waste waters.

21. Refractory Performance	Improved operating techniques better suited to the DOE-specific processing conditions are required for long-term processing success.
22. Nitrate Destruction	Methods are needed to destroy or remove residual nitrates from sludges and wastewaters.
23. Sludge Washing	Approaches are required to enhance the performance of candidate sludge washing technologies to demonstrate feed preparation and washing of process residues, sludges and particulates to RCRA requirements.
24. Molten Product Decanting	Operating techniques and equipment design are required to facilitate decanting or transfer of molten materials from furnaces in an effective, reliable and safe manner in a radioactive environment.

Source: DOE, 1997a

APPENDIX D

Biographical Sketches of Committee Members

Dejonghe, Paul A., *Chair*, received a B.S. in Agricultural Chemical Engineering and a Ph.D. in Applied Science from the University of Gent, Belgium. He is currently the consultant/advisor to the president at the Nuclear Energy Research Center (CEN/SCK) Mol, Belgium, and emeritus professor of the Faculty Applied Sciences of the Leuven University. He held several positions while at CEN/ SCK, including division head, and assistant general manager. Dr. Dejonghe has been involved with various committee studies on radioactive waste management of the IAEA; chairman of the Radioactive Waste Management Committee, OECD/NEA; past chairman of the Program Committee on Radioactive Waste Management, and the Committee on Plan of Action: Radioactive Waste Management, CEC. He has been a member of the National Research Council's Committee on Environmental Management Technologies since 1994 and chairman of the Committee on Mixed Waste since 1995.

Clarke, Ann N., holds a B.S. in Chemistry from Drexel Institute of Technology, and M.A.s in Chemistry and Earth Sciences from Johns Hopkins University; and a Ph.D. in Chemistry from Vanderbilt University. She was the Director of the Remediation Technologies Development Division at Eckenfelder, Inc., in Tennessee and is currently the President of ANC Associates, Inc. Prior to joining Eckenfelder, Dr. Clarke served as a Research Chemist for the U.S. Department of Agriculture, a Consulting Engineer at Sheppart T. Powell and Associates, and as an adjunct assistant professor at Vanderbilt University. She has experience in in-situ/ex-situ waste treatment research and in the design and implementation of projects involving alternative and innovative technologies for hazardous waste management. Dr. Clarke is a member of several professional societies, including the American Chemical Society, International Association on Water Pollution Research and Control, and Sigma Xi. Dr. Clarke has also won the Educator of the Year award from the National Environmental Training Association.

Exner, Jurgen H., received his B.S. from the University of Minnesota and his Ph.D. in Physical Organic Chemistry from the University of Washington. Dr. Exner is Principal and President of JHE Technology Systems, Inc., a consulting company specializing in waste management technology commercialization and application. He has experience in assessing environmental information and developing effective solutions based on regulatory, economic, technical, social, and legal considerations. His expertise lies in waste treatment and management, site investigation and feasibility studies, remediation, and in the application of thermal, chemical, physical, and biological treatment methods to solve environmental problems. Dr. Exner has 22 years of experience in hazardous waste management. He is associate editor of the *Air & Waste Management Association Journal* and chair of the Division of Environmental Chemistry of the American Chemical Society.

Hansen, Kent F., received his S.B. in Physics and Sc.D. in Nuclear Engineering from the Massachusetts Institute of Technology. With the exception of a short period in industry, Dr. Hansen has spent his professional career at MIT where he is currently a professor in the Department of Nuclear Engineering. His professional interests include numerical analysis, computational methods, modeling of engineering, managerial, and social systems; nuclear reactor theory and mathematics, reactor safety analysis, nuclear fuel management; engineering education, and energy systems. Dr. Hansen has been a consultant to major U.S. companies, four national laboratories, four government departments, and to the Electricite' de France. He is also a fellow of the Sigma Xi and American Nuclear Society. Dr. Hansen received the Arthur Holly Compton Award and became a member of the National Academy of

Engineering in 1982. He served as Director of the American Nuclear Society and is currently a director of EG&G, Inc., and Stone and Webster, Inc.

Lighty, JoAnn S., holds a B.S. and Ph.D. in Chemical Engineering from the University of Utah. She is currently an Associate Professor in the Department of Chemical and Fuels Engineering at the University of Utah. Dr. Lighty is an expert on high temperature processes. Her research includes biomass combustion, fate of metals during incineration, and characterization of particulate matter from combustion sources. Dr. Lighty is a member of the Society of Women Engineers, the Combustion Institute, and the American Institute of Chemical Engineers.

Samelson, Richard J., received a B.S. in Chemical Engineering from Iowa State University. After 40 years of service, he retired from PPG Industries in 1994. While at PPG, he worked within the Chemicals Group as manager and then director of Environmental Programs, manager of MIS Technical Support, manufacturing engineer for Inorganic Chemicals, R&D engineer and plant process engineer. Mr. Samelson's later responsibilities included environmental management and control, risk evaluation, and management of projects for the investigation and control of air and water pollution associated with past waste disposal practices. During his career, he served as a member of the Chemical Manufacturers Association Environmental Management Committee, and the Environmental Protection Committee of the Chlorine Institute, where he served two years as vice chairman and chairman respectively.

Steindler, Martin J., received his B.S., M.S., and Ph.D. in Chemistry from the University of Chicago. Dr. Steindler retired from the Argonne National Laboratory in 1993 as Director of Chemical Technology. He is a specialist in the nuclear fuel cycle and the related disciplines of nuclear chemistry and engineering. Dr. Steindler has been or is a consultant to various offices and departments of the AEC, ERDA, and DOE and to many DOE Laboratories, including Los Alamos, ORNL, Hanford, the Enrichment Technology Department at K-25, Rocky Flats, Livermore, and Savannah River. He was chairman of the Materials Review Board for the DOE Office of Civilian Radioactive Waste Management. He was a member of the Technology and Engineering Review Group for the AVLIS program and also served as the Argonne representative to the LAGER Committee of the DOE. He served on the USNRC Atomic Safety and Licensing Board Panel for 18 years and as a consultant to the waste management subcommittee of the Advisory Committee on Reactor Safeguards starting in 1974. He was appointed to the Advisory Committee on Reactor Safeguards and when the USNRC formed its Advisory Committee on Nuclear Waste (ACNW) he served as its first vice chairman until 1993 and later as Chairman until 1996. Dr. Steindler is a member of the American Chemical Society, the American Nuclear Society, the Royal Society of Chemists, Sigma Xi, and the American Institute of Chemical Engineers.

Thomson, Bruce M., holds a B.S. in Civil Engineering from the University of California, Davis, and M.S. and Ph.D. degrees in Environmental Science and Engineering from Rice University. He is currently a professor of civil engineering at the University of New Mexico. Dr. Thomson has worked extensively in waste management of hazardous metals and radionuclides, and has published on contaminant transport, waste disposal, and remediation of these contaminants. He has worked on research projects and consulted with Sandia and Los Alamos National Laboratories. He has received research support from numerous federal and state agencies including the USEPA, DOE, Forest Service, and the Bureau of Mines. He serves on the City of Albuquerque's Ground Water Protection Advisory Board and the New Mexico State Underground Storage Tank Committee. He is a registered professional engineer and is a member of several professional societies including the American Chemical Society, the Water Environment Federation, the American Society of Civil Engineers, and the National Ground Water Association.

APPENDIX E

Acronyms and Definitions

AEA	Atomic Energy Act of 1954
ALT	Accelerated leach test
AMWTP	Advanced Mixed Waste Treatment Project
ASTM	American Society for Testing and Materials
BDAT	Best demonstrated achievable technology
BRWM	Board on Radioactive Waste Management
CEMT	Committee on Environmental Management Technologies
CERCLA	Comprehensive Environmental Response, Compensation and Liability Act
CFR	Code of Federal Regulations
DOE	U.S. Department of Energy
DOT	Department of Transportation
EM	DOE Office of Environmental Management

EPA	U.S. Environmental Protection Agency
FFCA	Federal Facility Compliance Act
HIC	High integrity container
HLW	High-level waste
HSWA	Hazardous and Solid Waste Act
HWIR	Hazardous Waste Identification Rule
INEEL	Idaho National Engineering and Environmental Laboratory
LDR	Land Disposal Restrictions
LLW	Low-level waste
MLLW	Mixed low-level waste
MTRU	Mixed transuranic waste
MW	Mixed waste
MWFA	Mixed waste focus area
MWIR	Mixed waste inventory report
NDT	Nondestructive testing
NRC	National Research Council
OST	Office of Science and Technology
PA	Performance assessment
PCB	Polychlorinated biphenyl

PE	Performance evaluation
RCRA	Resource Conservation and Recovery Act
STCG	Site Technology Coordinating Group
STP	Site treatment plan
TCLP	Toxicity Characteristic Leaching Procedure
TRU	Transuranic (waste)
TSCA	Toxic Substances Control Act
USNRC	U.S. Nuclear Regulatory Commission
UTS	Universal Treatment Standards
WAC	Waste acceptance criteria
WIPP	Waste Isolation Pilot Plant